Climate Change Impacts on Coastal Soil and Water Management

Climate Change Impacts on Coastal Soil and Water Management

Zied Haj-Amor
Salem Bouri

CRC Press
Taylor & Francis Group
Boca Raton London New York

CRC Press is an imprint of the
Taylor & Francis Group, an **informa** business

CRC Press
Taylor & Francis Group
6000 Broken Sound Parkway NW, Suite 300
Boca Raton, FL 33487-2742

© 2020 by Taylor & Francis Group, LLC

CRC Press is an imprint of Taylor & Francis Group, an Informa business

No claim to original U.S. Government works

Printed on acid-free paper

International Standard Book Number-13: 978-0-367-40553-3 (Hardback)

This book contains information obtained from authentic and highly regarded sources. Reasonable efforts have been made to publish reliable data and information, but the author and publisher cannot assume responsibility for the validity of all materials or the consequences of their use. The authors and publishers have attempted to trace the copyright holders of all material reproduced in this publication and apologize to copyright holders if permission to publish in this form has not been obtained. If any copyright material has not been acknowledged, please write and let us know so we may rectify in any future reprint.

Except as permitted under U.S. Copyright Law, no part of this book may be reprinted, reproduced, transmitted, or utilized in any form by any electronic, mechanical, or other means, now known or hereafter invented, including photocopying, microfilming, and recording, or in any information storage or retrieval system, without written permission from the publishers.

For permission to photocopy or use material electronically from this work, please access www.copyright.com (www.copyright.com/) or contact the Copyright Clearance Center, Inc. (CCC), 222 Rosewood Drive, Danvers, MA 01923, 978-750-8400. CCC is a not-for-profit organization that provides licenses and registration for a variety of users. For organizations that have been granted a photocopy license by the CCC, a separate system of payment has been arranged.

Trademark Notice: Product or corporate names may be trademarks or registered trademarks, and are used only for identification and explanation without intent to infringe.

Library of Congress Cataloging-in-Publication Data

Names: Haj-Amor, Zied, author. | Bouri, Salem, author.
Title: Climate change impacts on coastal soil and water management / by Zied Haj-Amor and Salem Bouri.
Description: First edition. | Boca Raton, FL : CRC Press/ Taylor & Francis Group, 2020.
Identifiers: LCCN 2019043868 (print) | LCCN 2019043869 (ebook) | ISBN 9780367405533 (hardback ; acid-free paper) | ISBN 9780429356667 (ebook)
Subjects: LCSH: Shore protection. | Coast changes. | Soil erosion–Climatic factors. | Coastal zone management. | Climatic changes.
Classification: LCC TC330 .H345 2020 (print) | LCC TC330 (ebook) | DDC 627/.58–dc23
LC record available at https://lccn.loc.gov/2019043868
LC ebook record available at https://lccn.loc.gov/2019043869

Visit the Taylor & Francis Web site at
www.taylorandfrancis.com

and the CRC Press Web site at
www.crcpress.com

Contents

List of Figures xi
List of Tables xiii
Preface xv
Authors xvii
Acknowledgments xix
Acronyms and Abbreviations xxi
Nomenclature and Units xxiii
General Introduction xxv

Chapter 1   Soil–Water–Climate Interactions: Consideration for Climate Change Research 1

    1.1 Introduction 1
    1.2 Soil Moisture Definitions 1
    1.3 Soil Moisture Measurements 6
    1.4 Soil Moisture–Climate Interactions under Climate Change Condition 10
    1.5 Conclusions 11

Chapter 2   Soil–Plant–Climate Interactions: Considerations for Climate Change Research 13

    2.1 Introduction 13
    2.2 Plant–Soil Interactions: Nutrient Uptake 13
    2.3 Nutrient Uptake Simulation 17
    2.4 Plant–Soil Interactions and Climate Change 18
    2.5 Conclusions 20

Chapter 3   Soil Processes and Soil Properties 23

    3.1 Introduction 23
    3.2 Soil Forming Factors and Processes 23
        3.2.1 Soil Forming Factors 23
        3.2.2 Soil Forming Processes 26
    3.3 Soil Properties 27
        3.3.1 Physical Soil Properties 27
        3.3.2 Chemical Soil Properties 30
        3.3.3 Biological Soil Properties 33
    3.4 Conclusions 33

Chapter 4	Aspects of Global Climate Change 35	
	4.1 Introduction.. 35	
	4.2 Aspects of Global Climate Change................................ 35	
	4.2.1 Increasing Air Temperature 35	
	4.2.2 Rainfall Patterns Change 38	
	4.2.3 Sea Level Rise... 39	
	4.2.4 Ocean Acidification... 40	
	4.2.5 Floods... 41	
	4.2.6 Droughts... 42	
	4.3 Conclusions 43	
Chapter 5	Impacts of Climate Change on Soil Resources 45	
	5.1 Introduction.. 45	
	5.2 Effects of Climate Change on Soil Properties............... 45	
	5.2.1 Effects on Physical Soil Properties 45	
	5.2.2 Effects on Chemical Soil Properties 48	
	5.2.3 Effects on Biological Soil Properties 50	
	5.3 Climate Change and Land Degradation Processes........ 51	
	5.3.1 Soil Erosion Processes.. 51	
	5.3.2 Soil Salinization 52	
	5.3.3 Soil Acidification 53	
	5.4 Conclusions 54	
Chapter 6	Impacts of Climate Change on Water Resources.......................... 55	
	6.1 Introduction.. 55	
	6.2 Effects of Climate Change on Water Resources Availability.. 55	
	6.3 Effects of Climate Change on Surface Water Quality.......... 57	
	6.4 Effects of Climate Change on Groundwater.................. 61	
	6.4.1 Effects of Climate Change on Groundwater Recharge and Storage 61	
	6.4.2 Effects of Climate Change on Groundwater Quality.. 63	
	6.4.3 Effects of Sea-Level Rise on Groundwater............... 65	
	6.5 Conclusions 66	
Chapter 7	Potential for Soils to Mitigate Climate Change 67	
	7.1 Introduction 67	
	7.2 Soil Carbon Sequestration to Mitigate Climate Change 67	
	7.2.1 Biological Sequestration 70	

Contents

		7.2.2	Geological Sequestration	70
		7.2.3	Mineral Sequestration	72
		7.2.4	Ocean Sequestration	76
	7.3	Recent Soil Management Practices for Increasing Carbon Storage in Soil		77
	7.4	Conclusions		81

Chapter 8 Climate System Modeling 83

    8.1   Introduction 83
    8.2   Components of Climate Models 83
           8.2.1   The Atmospheric General Circulation Model (AGCM) 84
           8.2.2   The Ocean Model 86
           8.2.3   The Land Model 87
           8.2.4   Sea Ice Model 89
    8.3   Numerical Resolution of the Basic Equations 90
    8.4   Input and Output Data 92
    8.5   Strengths and Limitations of Climate Modeling 94
    8.6   Conclusions 95

Chapter 9 Coastal Water Resources and Climate Change Effects 97

    9.1   Introduction 97
    9.2   Coastal Water Resources 97
    9.3   Classification of Coastal Aquifers 98
           9.3.1   Confined Aquifers 98
           9.3.2   Unconfined Aquifers 99
    9.4   Properties of Coastal Aquifers 100
           9.4.1   Porosity 100
           9.4.2   Hydraulic Head 101
           9.4.3   Hydraulic Gradient 102
           9.4.4   Hydraulic Conductivity 102
           9.4.5   Aquifer Transmissivity 103
           9.4.6   Storage Coefficient 103
    9.5   Estimation of Aquifer Properties 104
    9.6   Effects of Climate Change on Coastal Aquifers 110
           9.6.1   Effects of Climate Change on Aquifer Recharge 110
           9.6.2   Sea Level Rise and Saltwater Intrusion in Coastal Aquifers 111
    9.7   Conclusions 112

Chapter 10 Climate Change and Coastal Erosion.. 115

    10.1 Introduction.. 115
    10.2 Contributory Factors to Coastal Erosion 115
        10.2.1 Natural Factors and Coastal Erosion................... 115
        10.2.2 Human-Induced Factors and Coastal Erosion..... 119
    10.3 Erosion Modeling as a Tool for Assessing Climate
        Change Impacts 120
        10.3.1 Empirical Models.. 121
        10.3.2 Physically Based Models 121
        10.3.3 Hybrid Models... 122
    10.4 Adaption Strategies to Protect Coastal Regions from the
        Impacts of Erosion.. 122
    10.5 Conclusions 123

Chapter 11 Climate Change and Water Resources Quality in
Coastal Regions 125

    11.1 Introduction.. 125
    11.2 Impacts of Extreme Climate Events on Water Resources.. 125
    11.3 Impacts of Climate Change on Water Bodies..................... 127
        11.3.1 Impacts on Lakes and Reservoirs.......................... 127
        11.3.2 Impacts on Alpine Lakes 128
        11.3.3 Impacts on River Resources................................. 129
        11.3.4 Impacts on Coastal Lagoons and Estuaries 130
    11.4 Potential Strategies for Reducing the Impacts of
        Climate Change on Water Quality 131
    11.5 Conclusions 132

Chapter 12 Soil and Water Management Strategies for Climate Change
Adaptation in Coastal Regions 133

    12.1 Introduction.. 133
    12.2 Soil and Water Management Strategies.............................. 133
        12.2.1 Conservation of Key Functions of Soil and
            Water under Climate Change................................ 134
            12.2.1.1 Enhance Soil Health............................. 134
            12.2.1.2 Conserve Water Quality 137
        12.2.2 Managed Aquifer Recharge 139
            12.2.2.1 Mechanisms of Aquifer Recharge 140
            12.2.2.2 Factors Influencing Performance
                      of Aquifer Recharge 142
            12.2.2.3 Environmental Risks of Aquifer
                      Recharge Projects................................ 144

		12.2.3 Measures for Reducing Soil Erosion	147
	12.3	Conclusions	150

General Conclusions 151

References 153

Index 183

List of Figures

Figure 1.1	Various processes and feedback loops within the climate system (adapted from Cubasch et al. 2013)	2
Figure 1.2	Volumetric soil moisture in the soil (adapted from Han et al. 2015)	3
Figure 1.3	Measurement of soil moisture through Time Domain Reflectometry method	8
Figure 1.4	An overview of soil moisture measurement through remote sensing method	9
Figure 2.1	Absorption of water and mineral nutrients from the soil by plant root	14
Figure 2.2	Diagram of the global carbon cycle. Boxes show the approximate size of carbon stores (in gigatons) (adapted from Falkowski et al. 2000)	19
Figure 3.1	Soil formation factors	24
Figure 3.2	Typical horizons of soil profile	25
Figure 3.3	Common organisms that can live in and on the soil	26
Figure 3.4	Main soil components	28
Figure 3.5	Textural triangle showing soil textural classes	29
Figure 4.1	Warming of earth surface	36
Figure 4.2	Observed and projected global average temperature from 1901 to 1921	36
Figure 4.3	Estimates of the current global water budget and its annual flow using observations from 2002–2008 (adapted from Trenberth et al. 2011)	38
Figure 4.4	Global average sea level from 1990 to 2010 (adapted from IPCC 2013)	39
Figure 4.5	Ocean acidification process	40
Figure 4.6	Storm damage in Central Texas	41
Figure 4.7	A schematic diagram showing the drought propagation under climate change (adapted from Mukherjee et al. 2018)	43
Figure 5.1	Soil moisture–atmosphere feedbacks. The plus and minus indicate positive and negative feedback loops, respectively. Soil moisture–temperature (black), soil moisture–precipitation (dark grey) and soil moisture–radiation (light grey) feedback loops	46
Figure 5.2	Displacement of soil	51
Figure 5.3	Global distribution of salt-affected soil (adapted from Shahid et al. 2018.)	53
Figure 6.1	Influences of natural processes and human activities on hydrologic cycle	56
Figure 6.2	One of Japan's biggest dams	57

xi

Figure 6.3	Correlation between water temperature and dissolved oxygen concentration under different climate change scenarios: (a) S1 (b) S2 (c) S3 (d) S4 (e) S5 (f) S6 (g) S7 (h) S8 (i) S9 (details about these scenarios in Danladi Bello et al. 2017)	60
Figure 6.4	Impacts of precipitation and evapotranspiration on groundwater recharge (adapted from Manna et al. 2019)	62
Figure 7.1	Soil biological processes that affect consumption (a) or emissions (b) of greenhouse gases (adapted from Baldock et al., 2012)	68
Figure 7.2	An overview of carbon capture and storage	69
Figure 7.3	An overview of geological sequestration	71
Figure 7.4	An overview of mineral sequestration	73
Figure 7.5	Diagram showing a simplified representation of the Earth's annual carbon cycle (adapted from Courtesy of Oak Ridge National Laboratory, U.S. Dept. of Energy)	76
Figure 8.1	Main drivers of climate change (adapted from Cubasch et al., 2013)	84
Figure 8.2	A simplified representation of the different processes that have to be parameterized in the ocean model (adapted from Goosse et al., 2010)	86
Figure 8.3	Development, melt, and movement of sea ice	89
Figure 9.1	World map of coastal regions	98
Figure 9.2	Confined aquifers and unconfined aquifers	99
Figure 9.3	An overview of pumping test	105
Figure 9.4	Principle of electrical resistivity (ER) investigation	107
Figure 9.5	Principle of electromagnetic inductive technique (adapted from Beck 1981)	109
Figure 9.6	Saltwater intrusion into a coastal aquifer	112
Figure 10.1	Types of movement of soil due to wind action	116
Figure 10.2	Contribution of storms to coastal erosion	118
Figure 10.3	Breakwater in a coastal area	123
Figure 12.1	Windbreaks in North Dakota, USA (adapted from Wright and Stuhr 2002)	136
Figure 12.2	Main benefits of windbreaks in an agricultural field	137
Figure 12.3	Main components of managed aquifer recharge from the source to the end-use	139
Figure 12.4	Main mechanisms of aquifer recharge: recharge basins, injection wells, etc. (adapted from Fisher et al. 2017)	140
Figure 12.5	A schematic representation of cut-off drains (adapted from Hearn and Hunt 2011)	148
Figure 12.6	A photo of a Gabion wall (adapted from Chen and Tang 2011)	149

List of Tables

Table 1.1	Overview of Various Soil Moisture Measurements	6
Table 2.1	Mobile and Relatively Immobile Nutrients in Soil and in Plants	15
Table 2.2	Percent of Nutrient Uptake Possible (%) by a Corn Crop Supplied by Mass Flow, Diffusion, and Root Interception	17
Table 3.1	Diameter (size) of Sand, Silt, and Clay Particles	28
Table 3.2	Ionic Forms of Major Nutrient Elements	31
Table 3.3	Millequivalents of Common Exchangeable Cations and their Equivalent in ppm	32
Table 4.1	Contribution of Various Sectors to Greenhouse Gases Emissions	37
Table 6.1	Pathways That Climate Change Affects Components of Groundwater System	65
Table 7.1	Worldwide Storage Potential for CO_2	69
Table 7.2	Advantages and Disadvantages of the Dry Gas-Solid Reaction	74
Table 7.3	Maximum Allowable Reaction Temperatures at Corresponding Pressure for a Number of Minerals	74
Table 7.4	Carbonation Potential of Natural Minerals Through Direct Aqueous Mineral Carbonation	75
Table 7.5	Traditional and Recommended Soil Management	77
Table 8.1	Major Differences Between the Atmospheric Model and the Ocean Model	87
Table 8.2	Some Outputs of Climate Models	93
Table 9.1	Some Examples of Porosity of Rocks and Unconsolidated Sediments	101
Table 9.2	Typical Hydraulic Conductivity of Geological Units	103
Table 9.3	Aquifer Potential for Exploitation Based on Transmissivity Values	104
Table 9.4	Recommended Minimum Frequencies for Water Level Measurements for Pumping Test	106
Table 9.5	Typical Electrical Resistivity Values for Some Common Earth Materials	108
Table 11.1	Correlations of River Runoff and Major Ion Content of the Yellow River (China)	130
Table 12.1	Overview of Soil and Water Management Strategies Under Climate Change	134

Preface

The United Nations forsees global population growth will result in 9 billion people in 2050, the majority of whom will live in coastal regions. This population growth, together with ongoing migration from rural to urban areas means the coastal regions will face a great challenge to conserve available soil and water resources. Additionally, climate-change driven variations in mean sea level, wave conditions, storm surge, and river flows will inevitably have serious impacts on the coastal soil and water resources. Given these factors and the extremely high value of coastal regions worldwide, effective adaptation measures underpinned by reliable coastal climate change impacts assessments are essential to avoid massive future coastal zone losses. The present book aims to promote the adoption of best soil and water practices in the coastal regions by: discussing how climate change can affect coastal soil and water resources and suggesting suitable strategies of coastal soil and water resources to fight the negative impacts of climate change. Recommendations on specific soil and water use planning strategies to address climate change are provided that can be incorporated into national and international development plans. Methodologies are presented to implement these recommendations for adaptation to global climate change.

Authors

Dr. Zied Haj-Amor is a Researcher at Water, Energy and Environment Laboratory, National Engineering School of Sfax, Sfax University, Sfax City, Tunisia. Dr. Zied Haj-Amor has published over ten peer-reviewed papers in reputed international journals and conferences and has been the assistant project coordinator on many research projects from the Tunisian Government. The current project focuses on soil and water management under climate change in coastal oases of Tunisia. His current research interests include soil and water management, climate change, geo-informatics, and geographic information systems (GIS).

Prof. Salem Bouri is a Full Professor of Earth Sciences at Faculty of Sciences Sfax, Sfax University, Sfax City, Tunisia. Prof. Bouri has published over 100 peer-reviewed papers in reputed international journals and conferences, and has been the Principal Investigator on many research projects from the Tunisian Government. His current research interests include water management, climate change, geo-informatics, and hydrogeology.

Acknowledgments

Many colleagues from Sfax University, Tunisia have helped to improve the quality of the present book. We gratefully acknowledge, in particular, the contributions made by the colleagues of the Faculty of Sciences of Sfax (FSS) and National Engineering School of Sfax (ENIS). The author would like to thank Prof. Jean-Paul Lhomme (retired from the Laboratory for the Study of Interactions between Soil, Agrosystems and Water Systems, IRD, Montpellier, France) for his helpful comments. Also, the authors would like to express their thanks to Prof. Joseph Kloepper (from Auburn University, USA) for his language assistance.

Acronyms and Abbreviations

AC	Alternating Electric Current
AGCM	Atmospheric General Circulation Model
ANC	Acid Neutralizing Capacity
AOP	Advanced Oxidation Process
B	Boron
C	Condensation
Ca	Calcium
CEC	Cation Exchange Capacity
CH_4	Methane
Cl	Chlorine
CO_2	Carbon dioxide
CME	Coefficient of Model Efficiency
Cu	Copper
DOC	Dissolved Organic Carbon
E	Electric Intensity
EC	Electrical Conductivity
EM	Electromagnetic inductive technique
ER	Electrical Resistivity
E_s	Soil Evaporation
FDM	Finite Difference Method
Fe	Iron
FEM	Finite Element Method
GCMs	General Circulation Models
GIS	Geographic Information System
GPP	Gross primary production
INCA-SED	Integrated Catchment-Sediment model
IPCC	Intergovernmental Panel on Climate Change
K	Potassium
MAGIC	Model of Acidification of Groundwater and Catchment
Mg	Magnesium
Mn	Manganese
NBP	Net biome production
N	Nitrogen
NEP	Net ecosystem production
N_2O	Nitrous oxide
NPP	Net primary production
O_2	Dioxygen
P	Pressure
PDEs	Partial Differential Equations
P	Phosphorus
RE	Relative Error
RMSE	Root Mean Square Error

S	Sulfur
SWAT	Soil and Water Assessment Tools model
SWIM	Soil and Water Integrated Model
T	Temperature
TDR	Time Domain Reflectometry
USLE	Universal Soil Loss Equation

Nomenclature and Units

θ	Volumetric soil moisture ($m^3\ m^{-3}$)
θ_{SAT}	Saturation soil moisture ($m^3\ m^{-3}$)
θ_s	Saturation ratio (−)
θ_{FC}	Field capacity (mm)
θ_{WILT}	Wilting point (mm)
θ_a	Available water content in the soil (mm)
V_w	Water volume in the soil (m^3)
V_s	Soil volume (m^3)
V_{wsat}	Water volume in the soil under a saturation condition (m^3)
V_p	Pores volume in the soil (m^3)
S	Absolute soil moisture content (mm)
d	Total depth of root zone (mm)
Ψ_p	Pressure potential (KPa)
Ψ_m	Matric potential (KPa)
Ψ_g	Gravitational potential (KPa)
Ψ_o	Osmotic potential (KPa)
b	An empirical soil water retention parameter (−)
C_s	Salts concentration ($g\ m^{-3}$)
F	Diffusion rate (−)
D	Diffusion coefficient (−)
dc/dx	Concentration gradient (−)
c	Average nutrient concentration at the root surface ($mol\ m^{-3}$)
x	Distance (m)
t	Time (s)
$\vec{V}$	Velocity ($m\ s^{-1}$)
a	Object acceleration ($m\ s^{-2}$)
F	Net force applied on the object (N)
m	Mass of the object (Kg)
V_0	Initial velocity at time = 0 ($m\ s^{-1}$)
D_e	Effective diffusion coefficient of the solute (−)
C	Concentration of the solute in soil solution ($g\ m^{-3}$)
n	Soil porosity (−)
ρ	density of the fluid (−)
dV	Change in the volume (m^3)
dt	Change in the time (s)
dM	Change in the masse (Kg)
dT	Change in the temperature (°C)
dp	Change in the pressure (Pa)
dS	Change in the salinization ($g\ kg^{-1}$)
EC	Soil electrical conductivity ($mS\ m^{-1}$)
CEC	Cation exchange capacity (meq/100 g)
q	Specific humidity ($g\ kg^{-1}$)

R	Universal gas constant (= 8.31 J K^{-1})
F$_{sol}$	Absorption of solar radiation in the ocean (W m^{-2})
F$_{diff}$	Diffusion term (m^2.s^{-1})
R$_n$	Net radiation (W m^{-2});
H	Convective heat transfer between the soil surface layer and the atmosphere (W m^{-2})
LE	Latent heat due to evapotranspiration (W m^{-2})
G	Heat transfer from and to the ground (W m^{-2})
R$_{Surf}$	Surface runoff (mm)
R$_{Sub-Surf}$	Sub-surface runoff (mm)
ΔSM/Δt	Change in soil moisture over a time step (mm)
u	Drift velocity of the ice (m s^{-1})
f	Coriolis parameter (−)
τ_a	Wind stress on the ice (−)
τ_w	Water stress on the ice, respectively (−)
g	Acceleration due to gravity (= 9.81 m s^{-2})
μ	Sea surface height (m)
Q	Flow rate (m^3 s^1)
K	Hydraulic conductivity (m s^1)
A	Area of the aquifer (m^2)
h	Hydraulic head (m)
I	Hydraulic gradient (−)
Δh	Hydraulic head difference (m)
L	Distance between two points (m)
K$_i$	Intrinsic permeability (m s^{-1})
μ	Dynamic viscosity of water (8.90×10^{-4} Pa s^{-1})
T	Transmissivity (m^2 s^{-1})
b	Saturated thickness of the aquifer (m)
S	Storage coefficient (−)
ΔV	Electric potential difference (V)
I	Current intensity (A)
a	Electrode spacing (m)
ρ$_a$	Apparent electrical resistivity (Ω−m)
α	Soil erosion constant (−)
Z	Slope-gradient factor (−)
U	Vegetation-cover factor (−)
K	Soil erodibility (Mg h^{-1} MJ^{-1} mm^{-1})

General Introduction

Coastal regions are frontiers with 356,000 km globally (Central Intelligence Agency 2016). They cover 20% of the land on earth and contain an extensive tapestry of natural ecosystems. They are home to 50% of the global population and make an important contribution (> 60%) to international agricultural production due to their favorable biophysical and climatic conditions (Burke et al. 2001). As reported in Martinez et al. (2007), these coastal regions provide numerous environmental benefits. For example, the physical features of coastal ecosystems (e.g. reefs of mangrove) are vital for natural functions such as land accretion and help to control land erosion and other damage arising from wind and wave action. Furthermore, the coastal regions are used extensively and increasingly for several activities such as agriculture, trade, industry, and amenity (Visbeck et al. 2014).

In recent years, the sustainability of the coastal regions has been threatened by several serious problems such as land degradation, pollution of soil and water resources, urbanization, and intensification of industrial activities. These in turn affect human welfare through their effects on productivity, health, and amenity. Land degradation is considered as the most critical problem among the mentioned problems. Climate change, unsuitable agricultural water practices, and soil mismanagement are major causes of land degradation in the coastal regions. Monitoring of land degradation caused by these factors has been the subject of international scientific debates leading to a commitment of global decision makers to address these threats collectively.

Due to their localization (i.e. transition between land and sea) and climate change, coastal regions are negatively impacted by changes in rainfall and temperature patterns (IPCC 2013). Climate change can affect the coastal areas through direct and indirect ways. Direct effects include the progressive inundation from sea-level rise, heightened storm damage, loss of wetlands, and increased salinization from saltwater intrusion. Indirect effects occur through events such as river floods, pulses, and quality of runoff that originate off-site but that affect the coasts. Also, climate change may pose the greatest threat to agriculture by increasing the demand for water and available water supply, soil salinization and fertility, and crop yield (Karmakar et al. 2016).

It is important that greater attention be paid to understanding the contribution of climatic factors in land degradation, i.e. water and soil resources degradation. Responding to the challenges of climate change impacts on water and soil resources requires urgent adaptation strategies at the local, regional, national, and global levels. Coastal countries are being urged to improve and consolidate their water and soil resources management systems and to develop and implement practical measures, which have positive development outcomes that are resilient to climate change (IPCC 2013). This book discusses the experience of monitoring soil and water degradation in some coastal regions under current and climate change conditions. In the first chapters (Chapters 1–9), the recent understandings of the soil processes, soil properties, climate change aspects, and soil and water resources management are described and an in-depth analysis of the climate change impacts of soil and water resources is presented. In Chapters 10–12, key adaptation options are presented for minimizing the impact of climate change on soil and water resources in the coastal regions. Finally, the book concludes by outlining priorities for adapting strategies to mediate the effects of climate change on soil and water resources.

1 Soil–Water–Climate Interactions

Consideration for Climate Change Research

1.1 INTRODUCTION

The water stored in soil (i.e. soil moisture) is a key variable in soil–water–climate interactions. Various processes and feedback loops within the climate system (Figure 1.1) depend highly on this variable. It is the main contributor for the water and energy cycles. Through evaporation and plant transpiration process, stored water in soil plays an important role in controlling the exchange of water and heat energy between the soil and the atmosphere by affecting the development of weather patterns and the production of precipitation (Seneviratne et al. 2010).

In this chapter, we present the past understandings of the importance of soil moisture for the climate system and provide an analysis of the current perspectives for research in this field. Special attention is paid to the aspects of soil moisture–climate interactions under climate change condition.

1.2 SOIL MOISTURE DEFINITIONS

There are several definitions for soil moisture. Generally, soil moisture refers to the water contained in the unsaturated soil zone (Hillel 1998). The soil moisture in this zone (also called the vadose zone) represents the water that is available to crops (Figure 1.2).

Based on this general definition, the volumetric soil moisture (θ) in the soil volume can be determined through the following equation:

$$\theta = \frac{V_w}{V_s} \quad (1.1)$$

where θ is the volumetric soil moisture ($m^3\ m^{-3}$), V_w is the water volume in the soil (m^3), and V_s is the soil volume (m^3). In hydrological modelling, θ is expressed in mm_{H2O}/mm_{soil}. In this case, θ is commonly expressed as the amount of water (in mm of water depth) present in of the top one meter of soil ($mm\ m^{-1}$).

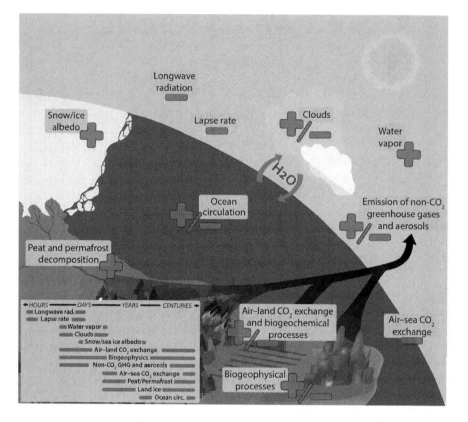

FIGURE 1.1 Various processes and feedback loops within the climate system (adapted from Cubasch et al. 2013.)

Due to the variability of the water content in the soil with time and space (i.e. as a function of rooting depth or the water table depth, Schenk and Jackson (2002), several soil moisture levels could be defined as follows:

- The saturation soil moisture (θ_{SAT}):

    Immediately after rain or irrigation events in an agricultural field, the soil pores will fill with water, and the soil is said to be saturated. Accordingly, the saturation soil moisture (θ_{SAT}) can be defined as the maximum water content that can be stored in the soil. The following equation can be used to calculate θ_{SAT}:

$$\theta_{sat} = \frac{V_{wsat}}{V_s} \quad (1.2)$$

where θ_{SAT} is the saturation soil moisture (m³ m⁻³), V_{wsat} is the water volume in the soil under a saturation condition (m³), and V_s is the

Soil–Water–Climate Interactions

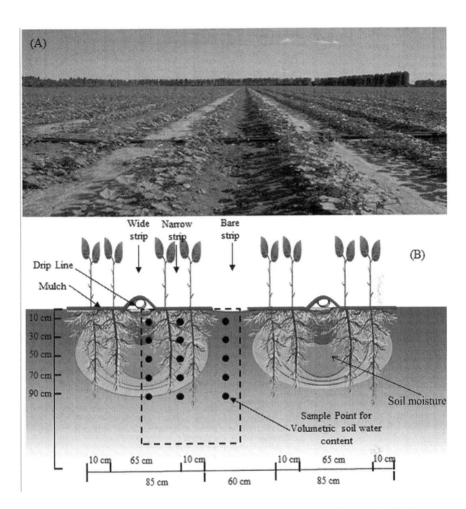

FIGURE 1.2 Volumetric soil moisture in the soil (adapted from Han et al. 2015.)

soil volume (m³). Physically, θ_{SAT} is the water content in the soil at − 1 hPa of hydraulic head. Using θ_{SAT}, we also can use the saturation ratio (Equation 1.3) to define the moisture level in the soil:

$$\theta_s = \frac{\theta}{\theta_{sat}} \qquad (1.3)$$

where θs ranges between 0 (dry soil) and 1 (saturated soil), and θ ranges between 0 and θ_{SAT}. The air and water contained in the soil are critical for crop growth. Under saturation conditions, no air is available, and the crops will suffer. Saturation condition for a long period (example: 3 days) may lead to serious stress on the crops. After rain or irrigation application, due to the drainage process, an important part of the water stored in the larger pores will move downward. The drained water will be replaced by the air which allowed to normal growth of crops. In coarse textured soils (example: sandy soils), drainage is completed within a period of a few hours. In fine textured soils (example: clayey soils), the drainage process may take 2 to 3 days (Hillel 1998).

- Field capacity (θ_{FC}):

Field capacity is the amount of water stored in the soil after drainage process which usually takes place 2 to 3 days after rain or irrigation applications. Alternatively, θ_{FC} can be defined as the threshold point at which the soil pore water will be influenced by drainage process. Physically, θ_{FC} is the water content in the soil at −33 kPa of hydraulic head (Dingman 2002).

- Wilting point (θ_{WILT}):

Wilting point is the minimal amount of water in the soil needed to avoid the wilting of plants. Accordingly, θ_{WILT} can be defined as the threshold point below which the plant roots will be unable to extract water from the soil. Physically, θ_{WILT} corresponding to soil moisture potential of −1500 kPa (Dingman 2002). θ_{WILT} depends on the crop type (Hupet et al. 2005). For example, temperate oaks can extract water at potentials of −2 MPa (Bréda et al. 1995), whereas, Mediterranean shrub species can extract water at potentials of −5 MPa (Rambal et al. 2003).

Using θ_{FC} and θ_{WILT}, the available water content in the soil (θ_a) which corresponds to the amount of water actually available to the plant, can be calculated as follows:

$$\theta_a = \theta_{FC} - \theta_{WILT} \tag{1.4}$$

θ_{FC}, θ_{WILT}, and θ_a are expressed in mm_{H2O}/mm_{soil}.

- The absolute soil moisture (S):

In addition to the relative soil moistures definitions (Equations 1.1 to 1.4), it is also possible to express the soil moisture in absolute terms by considering the total water content (θ) stored in the total depth of root zone (d). This is highly relevant to the computation of the terrestrial water balance (details about this balance in section 1.3). Accordingly, the absolute soil moisture content (S) is expressed in mm (Equation 1.5):

$$S = \theta * d \tag{1.5}$$

The soil moisture regime, determined by the changes in soil water content with time, is the main factor conditioning the plant growth and crop production. The above discussed soil moisture levels play a key role in the description of this regime.

Soil–Water–Climate Interactions

It is also useful to define the state of water in the soil as a function of its potential (Ψ). In soils, the driving force for water to flow is the gradient in total water potential. The total potential of bulk soil water can be written as the sum of all possible component potentials, so that the total water potential is equal to the sum of osmotic, matric, gravitational, and hydrostatic pressure potential. Whereas in physical chemistry the chemical potential of water is usually defined on a molar or mass basis, soil water potential is usually expressed with respect to a unit volume of water, thereby attaining units of pressure (Pa or kPa); or per unit weight of water, leading to soil water potential expressed by the equivalent height of a column of water (L). The resulting pressure head equivalent of the combined adsorptive and capillary forces in soils is defined as the matric pressure head, h. When expressed relative to the reference potential of free water, the water potential in unsaturated soils is negative (the soil water potential is less than the water potential of water at atmospheric pressure). Hence, the matric potential decreases or is more negative as the soil water content decreases. Often, Ψ is expressed in kPa (1 kPa = 1 J Kg^{-1} $\approx$ 0.1 m water column). Ψ is the sum of four components:

- Pressure potential (Ψ_p) is the pressure exerted by a water column above the point considered. Hence, under unsaturated conditions, $\Psi_p = 0$;
- Gravitational potential (Ψ_g) is the pressure of a water column resulting from its position in a gravitational field. Ψ_g is calculated as the product between the height above the arbitrary reference plane (h) and the gravitational constant (g = 9.81 m s^{-2});
- Matric potential (Ψ_m) is the pressure resulting from the attraction of the soil matrix and the water molecules. The "soil–water characteristic curve" is typically used to define the relationship between the soil water content and the potential matrix. Under saturated conditions, the potential matrix is equal to 0 and becomes more negative under drier conditions. In order to compute the soil–water characteristic curve, a simple function (Equation 1.6) was developed by Campbell (1974) as follows:

$$\psi_e = \psi_g \left(\frac{\theta_v}{\theta_{sat}}\right)^{-b} \quad (1.6)$$

where Ψ_m is the air-entry matric potential (cm H2O), θ_v is the volumetric soil water content, θ_{sat} is the volumetric soil water content under saturation condition, and b is an empirical soil water retention parameter, which depends mainly on soil texture (Clapp and Hornberger 1978);
- Osmotic potential (Ψ_o) is the pressure due to the presence of the salts in the soil solution. Hence, $\Psi_o = 0$ for water without salts and becomes more negative as the sum of salts in the soil solution increases. The relationship between Ψ_o (kPa) and the salts concentration (C$_s$, g m^{-3}) is described by the following quation:

$$\psi_o = -0.05625 \times C_s \tag{1.7}$$

At the end of this chapter (in Appendix 1.1), an exercise (with correction) about how to calculate the soil-water characteristic curve is presented.

1.3 SOIL MOISTURE MEASUREMENTS

Soil moisture data can be exploited in several ways. For example, the data can be used for reservoir management, droughts warning, irrigation scheduling, climate system characterization, and crop yield forecasting, etc. The soil moisture can be measured based on the following three methods: gravimetric method; neutron method; gamma attenuation; Time Domain Reflectometry; tensiometer method; gypsum block method; and hygrometric techniques. An overview of these various methods was presented in Table 1.1.

Depending on soil physical properties and the main objective of the soil moisture measurement, some methods are more effective than others (Munoz-Carpena et al. 2004). Furthermore, details about these methods were reported in Sharma et al. (2018) as follows.

- The gravimetric method:

    With this laboratory method, soil moisture is expressed as the ratio of the weight of water to the weight of oven dried soil (Reynolds 1970).

TABLE 1.1
Overview of Various Soil Moisture Measurements

Measurement Method	Measured Parameter	Advantages	Disadvantages
Gravimetric method	Water content mass	High level of accuracy So easy method	Destructive
Neutron method	Water content volume	Nondestructive High level of accuracy	Time intensive
Gamma attenuation	Water content volume	Nondestructive Automated method	Low level of accuracy
Time domain Reflectometry	Water content volume	Nondestructive So easy method	Sensitive to saline soil
Tensiometer method	Soil water potential	Nondestructive Easy installation	Automated operation impractical
Gypsum block method	Soil moisture tension	Nondestructive High level of accuracy	Calibration changes with time
Hygrometric technique	Soil water potential	Nondestructive Automated method	High interaction with soil

After soil sampling from the investigated field, a small soil sample is weighed under wet condition, oven dried (221 °F, for 2 hours), and reweighed after drying. Then, the water weight is calculated as the difference between the weight of the wet soil and the weight of oven dried soil. It is important to mention that this method is destructive and hence, not well suitable for continuous investigation, or when a high number of soil moisture observations is needed. Therefore, there is need to measure the soil moisture through alternative methods. Generally, these methods are based on the high relative permittivity of water in comparison to air;

- The neutron method:

This is based on the slowing-down by the water of fast neutrons emitted by a radioactive source. The soil moisture measurement is made by lowering a probe consisting of the source and a detector to the investigated soil depth in a suitable access hole in the ground. The soil moisture can be measured by a neutron probe if it is properly calibrated by gravimetric sampling (Greacen 1981);

- The gamma attenuation method:

This method has the advantage of being non-destructive, i.e. there is no need for direct contact between the investigated soil and the detector. But, it has also some disadvantages, in particular, the small depth of investigated soil (just a few centimeters).

The moisture can have a serious impact on the transmission of terrestrial gamma radiation through soil. Therefore, the moisture can affect the gamma radiation dose rate above the surface of the earth. Based on this assumption, equations connecting gamma ray counting rates with soil moisture were developed (details in Udagani 2013) and used to calculate the soil water content through the gamma attenuation method;

- The time domain reflectometry method:

This indirect measure of soil moisture is based on the principle of the dielectric constant of water which is much larger than that of other matter such as soil and air (Schanz et al. 2011). In any desired soil depth, soil moisture is measured by a sensor composed of two parallel rods attached to a coaxial cable with the same impedance as the output connection of the TDR probe (Figure 1.3);

- The tensiometer method:

The tensiometer is an instrument used to measure the potential of water content in the vadose zone (i.e. capillary tension) over the range from 0 to 1 atmosphere. The main components of this instrument are a sealed, water-filled tube with a ceramic porous cup on one end, and a gauge for measuring pressure. The porous cup is permeable to water (not to the soil matrix or to gases). Water moves through the cup until the water pressure

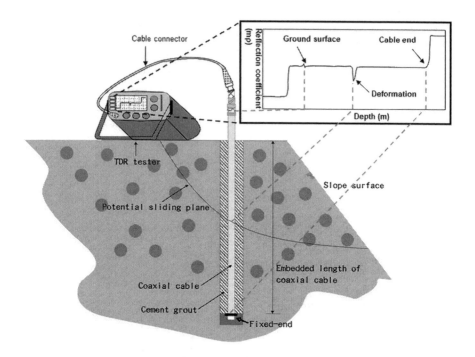

FIGURE 1.3 Measurement of soil moisture through Time Domain Reflectometry method.

inside the tensiometer is equal to the capillary tension. At equilibrium, the water pressure in the tensiometer is equal in magnitude to the soil matric potential (Munoz-Carpena et al. 2004);

- The gypsum block method:

  This method is usually used to measure soil water tension under saline conditions (i.e. saline soils) over the range from 30 to 1500 kPa. A gypsum block (cylindrical in shape) consists of two electrodes embedded in a block of gypsum. The gypsum block is an electrochemical cell with a saturated solution of calcium sulfate. The soil moisture is identified by measuring the resistance between these two electrodes. Wires are joined onto each electrode and extruded from the gypsum block to measure the resistance between the electrodes. The resistance between the two electrodes varies with the water content in the gypsum block, which will depend directly on the soil water tension. Under saline conditions, using this method may allow for some errors, hence, certain corrections functions are required. See Hoseini and Albaji (2016) for more details about these functions;

- The hygrometric method:

  This method is based on the thermal inertia of porous material which varies according to soil moisture. In this method, the electrical resistance hydrometer (i.e. an instrument) is used to measure the relative humidity of soil based on chemical salts and acid, aluminum oxide, electrolysis, thermal principles, and white hydrosol (Sharma et al. 2018).

A major disadvantage of the soil moisture measurement methods discussed above is that they are usually used to determine soil moisture in a specific point in small areas (a few m^2). Hence, they are rarely used to evaluate spatial distribution of soil moisture at larger scale (e.g. a few km^2). Evaluating soil moisture over a large agricultural area is critical for selecting suitable agricultural water management systems. Furthermore, evaluating soil moisture is critical for the assessment of key hydrological processes, including infiltration, soil erosion, and runoff. In this context, several alternative methods were adapted to help assess the spatial distribution of soil moisture on a larger scale. The remote sensing method (Figure 1.4) is considered as the commonly used method among these alternative methods (Engman and Chauhan 1995).

Measuring soil moisture by the remote sensing method began in the mid 1970s, immediately after the surge in satellite development. This method can estimate soil moisture over a great land surface; however, it has also some disadvantages. Only the moisture in the top few centimeters of soil can be estimated, and it is an indirect measure of soil moisture based on the principle of the large contrast between the dielectric properties of wet and dry soil. Quantitative measurements of soil moisture in the top soil layer have been most successful using passive remote sensing in the microwave region. The data obtained from the

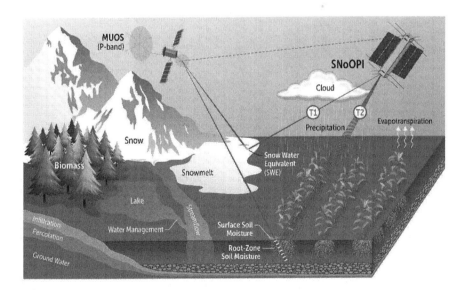

FIGURE 1.4 An overview of soil moisture measurement through remote sensing method.

microwave remote sensing satellites (e.g. WindSat, RADARSAT, and SMAP) were exploited to determine the soil moisture (Quesney et al. 2000).

1.4 SOIL MOISTURE–CLIMATE INTERACTIONS UNDER CLIMATE CHANGE CONDITION

Climate change is mainly attributed to anthropogenic activities that add large amounts of greenhouse gases (especially carbon dioxide (CO_2), methane (CH_4), and nitrous oxide (N_2O)) to those naturally presenting in the atmosphere. During the last century, changes in precipitation and temperature have been observed as a consequence of the large-scale emissions of greenhouse gases. Atmosphere is one of the key components of the climate system, and as the climate changes, characteristics of soil moisture–atmosphere coupling will likely change (Hirschi et al. 2011). Temperature and precipitation constitute the main elements of the atmosphere. Accordingly, in this section, we discuss characteristics of soil moisture–temperature coupling and then characteristics of soil moisture–precipitation coupling.

i. Soil moisture–temperature coupling:

As revealed by several modeling studies (e.g. Fischer et al. 2007; Mueller and Seneviratne 2012), the significant effect of soil moisture on near-surface climate is attributed in high part to the change in temperature: soil moisture controls the amount of latent and sensible heat flux. Low values of soil moisture lead to a larger proportion of sensible heat compared to latent heat. This in turn induces an increase of near-surface air temperature. Soil moisture–temperature interactions can significantly impact near-surface climate over the entire range of temperature, but, they have the most significant impact for extreme hot temperatures and heat waves (Seneviratne et al. 2006; Fischer et al. 2007; Zhang et al. 2009). The soil moisture–temperature coupling is expected to be stronger in transition zones between wet and dry climates (Santanello et al. 2011).

ii. Soil moisture–precipitation coupling:

As revealed by several mechanistic studies (e.g. Alfieri et al. 2008), the impact of soil moisture anomalies on boundary-layer stability and precipitation formation is mainly attributed to the various soil moisture–precipitation interactions. For example, the additional precipitated water falling over wet soils may originate from oceanic sources, but the triggering of precipitation may itself be the result of enhanced instability induced by the wet soil conditions (or in some cases dry soil conditions). As suggested by several studies (e.g. Koster et al. 2003; Alfieri et al. 2008), due to the complexity of soil moisture–precipitation feedbacks, direct analyses of soil moisture–precipitation feedbacks using soil moisture and precipitation observations coupled with the data of modelling tools are so useful for identifying regions of strong soil moisture–precipitation coupling.

1.5 CONCLUSIONS

In this chapter, we defined in detail the soil moisture term and its importance for the climate system, and provided an analysis of the current perspectives for research in this field. Also, important attention is paid to the aspects of soil moisture–climate interactions under climate change condition. The investigation of these aspects is a critical research field at the interface between many scientific disciplines (e.g. climate science, hydrology, atmospheric science, meteorology, soil physics, etc.). Therefore, combining all these disciplines would be useful to deepen our knowledge on the soil moisture–climate interactions worldwide. Further future work should combine and compare modeling and observational datasets, and should include conducting of numerical and field experiments.

APPENDIX 1.1 (EXERCISE WITH CORRECTION)

EXERCISE:

For a sandy loam soil, the soil–water characteristic curve is determined as follows: $\psi_m = -0.152\, \theta^{-3.1}$ (KPa)

1. Calculate the water potential for this soil at 1 m depth (water content = 0.2 m^3 m^{-3}; salts concentration = 64 g m^{-3});
2. A water table is located at a depth of 1 m in the investigated sandy loam soil. What would be the water content at the soil surface under hydraulic equilibrium condition if the osmotic potential does not change with depth?

CORRECTION:

1. We fix the reference level on the soil surface. So:

$$\Psi = \Psi_m + \Psi_g + \Psi_o = (-0.152 \times 0.2^{-3.1}) + (-9.81 \times 1) + (-0.05625 \times 64)$$
$$= (-22.4) + (-9.81) + (-3.6) = -35.7 \text{ KPa} = -0.036 \text{ MPa} = -.036 \text{ bar}.$$

2. If we take the soil surface as a reference level, the water potential at the water table will be: $\Psi = \Psi_m + \Psi_g + \Psi_o = 0 + + (-9.81 \times 1) + \Psi_o = -9.81 + \Psi_o$ (expressed in kPa).

    The water potential at the soil surface will have the same value. If osmotic potential is the same at the reference level and at the water table, the matric potential at the surface soil will be: $\Psi_{m\,surface} = -9.81 + \Psi_{o\,water\,table} - \Psi_{o\,surface} = -9.81$ KPa.

2 Soil–Plant–Climate Interactions

Considerations for Climate Change Research

2.1 INTRODUCTION

Soil provides several beneficial functions for plants. In addition to physical support, the soil is the main reservoir for water, mineral nutrients, and carbon. These elements play a key role in plant growth (Hell and Hillebrand 2001). Furthermore, water, mineral nutrients, and carbon determine plant health. Plants also change these resources via production of organic material, resource allocation belowground, and interactions with soil microorganisms. These processes have significant effects on soil structure, nutrient availability, and the water, carbon, and nitrogen cycles of soil, which are all linked to and feed back to plant growth, community composition, and resilience to climate stress (Connolly and Walker 2008). These processes are the result of various interactions between plant, climate, and soil. Understanding these processes and interactions is critical for evaluating plant nutrient relations (Schuur et al. 2009).

Changes in the climate and atmosphere (e.g. rising atmospheric CO_2 concentrations) have the ability to alter soil–plant interactions and their feedback to ecosystem functioning. For example, climate change has the ability to decrease the concentration of mineral nutrients available in the soil. Therefore, under continual global climate change condition, it is essential to understand the strategies that plants have evolved to cope with some of these obstacles. In this chapter, we evaluate the interactions between soil and plant, and discuss how these plant–soil interactions feed back to plant resilience under climate change, which could open new ways of predicting vegetation dynamics and mitigating the impacts of climate change on vegetation.

2.2 PLANT–SOIL INTERACTIONS: NUTRIENT UPTAKE

For a normal growth, the plants need the following elements from the soil: water, carbon dioxide, and mineral nutrients (such as N, P, K, Ca, Mg, microelements). These nutrients are presented in gaseous, liquid, and solid forms.

Generally, the mineral nutrients are classified in three categories: primary nutrients (nitrogen (N), phosphorus (P), potassium (K)); secondary nutrients (calcium (Ca), magnesium (Mg), sulfur (S)); and micronutrients (iron (Fe), boron (B), copper (Cu), chlorine (Cl), manganese (Mn), molybdenum (Mo), zinc (Zn)). Nutrient uptake of plants from soil is the result of interactions between plants and soil (plant–soil interactions) (Oates and Barber 1987).

Nutrients must come in contact with the root surface in order to be absorbed by the root (Figure 2.1).

The nutrients move to the plant root surface through three process: (1) mass flow; (2) diffusion; and (3) root interception. Mass flow and diffusion contribute significantly to this movement. However, several factors (e.g. high clay content and low soil water content) can have negative impacts on these two processes. It is important to note that the mobility of mineral nutrients in plants is different than their mobility in soil (Table 2.1).

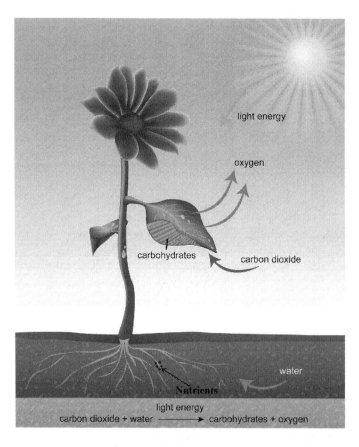

FIGURE 2.1 Absorption of water and mineral nutrients from the soil by plant root.

TABLE 2.1
Mobile and Relatively Immobile Nutrients in Soil and in Plants

Plants		Soil	
Mobile	Relatively Immobile	Mobile	Relatively Immobile
Chloride	Boron	H_3BO_3	NH_4+
Magnesium	Calcium	H_2BO_3	Ca^{+2}
Molybdenum	Copper	Cl^-	Cu^{+2}
Nitrogen	Iron	NO_3^-	Fe^{+2}, Fe^{+3}
Phosphorus	Manganese	SO_4^{-2}	Mg^{+2}

As mentioned in Table 2.1, the list of mobile and immobile nutrients is somewhat different in soil than in the plant, yet is very important in exploring why some mineral nutrients limit growth more than others. Specifically, nutrient mobility affects fertilization practices. For example, N fertilizer can be broadcast or incorporated with fairly similar results because it is quite mobile. However, P fertilizer is generally either banded or applied with the seed because it is quite immobile in most soils.

(1) Mass flow process:

Mass flow is a convective process in which dissolved nutrients and other substances are moved in the flow of water to the plant roots due to transpiration phenomenon. Mass flow of dissolved nutrients from the roots to the leaves is driven in great part by water potential differences and also by capillary action. Details about water potential were presented in Chapter 1. Sometimes, percolation of soil water can also contribute to minor mass flow (Roose and Kirk 2009). Also, all of the water in soil is not used by the plant. Indeed, the soil water contains three negatively charged ions (nitrogen, sulfur, and boron). Likewise, all of the negatively charged ions in this water will not be used by plants. Some of the water and the nutrients contained in this water move below the root zone due to leaching process (Schenk and Barber 1980). The quantity of the dissolved nutrients that are moved to the plant roots due to the mass flow process is not a constant value. Many factors can affect mass flow, including the soil volume occupied by the roots, the concentration of nutrients in the soil, and root morphology (Samal et al. 2010). Movement of ions in the soil solution to the roots contributes significantly to plant nutritional needs.

(2) Diffusion process:

Diffusion is the process of movement of nutrient ions from high to low concentration area by random thermal motion (Crank 1956). This movement is assured only when the concentration at the root surface is either higher or lower than that of the surrounding solution. Under insignificant transpiration of plants and high concentration of ions in soil solution, the diffusion process plays a significant role for nutrient uptake of plant root from soil. Under reverse conditions, it is the mass flow process that provides this significant role. The diffusion process is driven by a gradient in concentration. As plant roots absorb nutrients from the surrounding soil solution, the diffusion gradient is set up. A high plant requirement or a high root "absorbing power" results in a strong sink or a high diffusion gradient favoring ion transport (Kirkham 1994). The diffusion process has a close relationship with the photosynthesis process. Indeed, it is the underlying process for photosynthesis where CO_2 from stomata diffuses into the leaves and then to the cells. In contrast, by transpiration the diffusion of water and O_2 happens into the environment (Sallam et al. 1984). Several factors have significant impacts on the diffusion process. However, the most significant factors are: (1) the diffusion coefficient (i.e. a parameter expressing the transfer rate of a nutrient ion by random thermal motion); (2) the concentration of the nutrient in the soil solution; and (3) the buffering capacity of the solid phase of the soil for the nutrient in the soil solution phase (Kirkham 1994). The diffusion rate (quantity diffused per unit cross section and per unit time) can be determined based on Fick's equation as follows:

$$F = -D(dc/dx) \qquad (2.1)$$

where F is the diffusion rate, D is the diffusion coefficient, dc/dx is the concentration gradient, c is the average nutrient concentration at the root surface, and x is the distance. Several factors can affect the diffusion rate such as the gradient of concentration, the temperature, and the permeability of membranes.

Although the mineral nutrients (especially phosphorus and potassium) are highly absorbed strongly by soils, they show low concentration in the soil solution.

Usually, these nutrients (with low concentration) move to the plant roots by the diffusion process. As uptake of these mineral nutrients occurs at the plant roots, the concentration in the soil solution in close proximity to the root decreases. This creates a gradient for the nutrient to diffuse through the soil solution from a zone of high concentration to the depleted solution adjacent to the plant roots. Diffusion is responsible for the majority of the P, K, and Zn moving to the plant roots for uptake.

(3) Root interception:

This is the movement of roots through the soil and the absorption of nutrients as the roots come in contact with nutrients (Dexter 1987). Several factors can enhance the nutrient uptake by this movement.

Soil–Plant–Climate Interactions

The main factors are the growth of new roots throughout the soil mass, the good soil management, and the mycorrhizal infections. Indeed, as the root system grows and exploits the soil more completely, soil solution and soil surfaces retaining adsorbed ions are exposed to the root mass and absorption of these ions by the contact exchange is accomplished (Czarnes et al. 2000).

At the end of this section, in Table 2.2, an example shows the contribution of each of the three processes discussed above to the total supply of the nutrients for the corn crop. Furthermore, at the end of this chapter (in Appendix 2.1), an exercise (with correction) about the nutrient uptake is presented.

2.3 NUTRIENT UPTAKE SIMULATION

Simulation of nutrient uptake is essential to predicting plant growth under nutrient limitation (Ptashnyk 2010). Usually, the simulation of the rate of the diffusive mass transport is based on "Governing Partial Differential Equation" (Equation 2.2), which was developed based on a combination between Fick's equation and the law of mass conservation (Kirkham and Powers 1972). In this equation, solute diffusion in a soil medium with a constant water content and a diffusion coefficient is determined as follows:

$$\theta \frac{\partial C}{\partial t} = D_e \frac{\partial^2 C}{\partial x^2} \tag{2.2}$$

where C is the concentration of the solute in soil solution, x is the position (or distance), t is the time, θ is the volumetric water content, and D_e is the effective diffusion coefficient of the solute. D_e can be determined from the solute's diffusion coefficient in water (D_0), soil porosity (water content), and soil water content (θ_s) based on the equation developed by Millington and Quirk (1961) (Equation 2.3):

TABLE 2.2

Percent of Nutrient Uptake Possible (%) by a Corn Crop Supplied by Mass Flow, Diffusion, and Root Interception (Adapted from Mengel 1995)

Nutrient	Mass Flow (%)	Diffusion (%)	Root Interception (%)
Nitrogen (N)	80	19	1
Phosphorus (P)	5	93	2
Potassium (K)	18	80	2
Calcium (Ca)	375	0	150
Magnesium (Mg)	600	0	33
Sulfur (S)	300	0	5

$$D_e = D_0 \frac{\theta^{10/3}}{\theta_s^2} \qquad (2.3)$$

The resolution of the Governing Partial Differential Equation requires defining the initial and boundary conditions of soil medium. Usually, these conditions are defined as follows: the diffusing substance has an initial concentration C_o in soil water from position 0 to position h; The diffusing substance has an initial concentration of zero at other positions. At times greater than 0, the material diffuses through the soil, but none of the material diffuses out of the soil at position 0 (Equation 2.4):

$$\begin{cases} C(x,0) = C_0 \text{ for } 0 < x < h \\ C(x,0) = 0 \text{ for } x \geq h \\ \frac{\partial C}{\partial x} = 0 \text{ for } x = 0 \text{ and } t \geq 0 \end{cases} \qquad (2.4)$$

According to these initial and boundary conditions, the solution to the Governing Partial Differential Equation is as follows (Equation 2.5):

$$C(x,t) = \frac{1}{2} C_0 \left\{ erf \left[\frac{h-x}{2\sqrt{(De/\theta)t}} \right] + erf \left[\frac{h+x}{2\sqrt{(De/\theta)t}} \right] \right\} \qquad (2.5)$$

where erf is the error function. This function is determined based on the x distance.

Based on the equations presented above (i.e. Equations 2.2, 2.3, 2.4, and 2.5), several simulation models have been developed and have been extensively used during the last decade to investigate the nutrient transfer mechanisms and to define the soil management option for the optimization of crop nutrition and sustainable systems (Roose et al. 2001). Studying the nutrient transfer mechanisms has been one of the very active areas of soil fertility research in recent years (Ptashnyk 2010). Nutrient uptake models proved to be an effective tool in studying the processes that govern soil supply and plant uptake of essential nutrients. For example, these developed models have been used in evaluating the effects of changing soil bulk density on K^+ uptake by soybeans, in investigating the impact of soil-related factors governing K^+ availability on K^+ uptake by corn, in studying the impact of sewage sludge addition on Zn uptake in investigating the impact of soil pH on phosphorus and potassium uptake, in quantifying the amount of phosphate released from rock phosphate by rape.

2.4 PLANT–SOIL INTERACTIONS AND CLIMATE CHANGE

As revealed by several international studies (Heath et al. 2005; Steinbeiss et al. 2008), the transfer of carbon through plant roots to the soil has a positive role in regulating ecosystem responses to climate change and its mitigation. In this

Soil–Plant–Climate Interactions 19

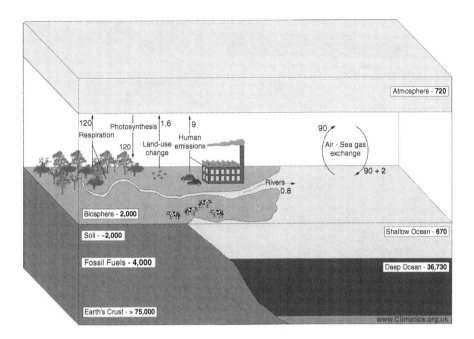

FIGURE 2.2 Diagram of the global carbon cycle. Boxes show the approximate size of carbon stores (in gigatons) (adapted from Falkowski et al. 2000.)

section, we study the mechanisms involved in this transfer, the consequences for ecosystem carbon cycling, and we investigate the potential to exploit plant root traits and soil microbial processes that favor soil carbon sequestration.

Through carbon decomposition and a range of additional biologically mediated transformations, the soil has the potential to absorb a great quantity of carbon (Figure 2.2). Therefore, this great reservoir (i.e. soil) will enhance mitigatation of climate change through its potential to sequester carbon from the atmosphere (De Deyn et al. 2011). This potential, often called "gain of carbon in soil" process, is highly regulated by plant–microbial–soil interactions.

Climate change can affect soil carbon through direct and indirect ways. Considering the direct effects, a subtle warming (by 1°C for example) can directly stimulate microbial activity allowing for a significant increase in ecosystem respiration rates in arable lands. Considering the indirect effects, it is so important to note that climate change has a significant indirect effect on soil microbes and carbon cycling, by affecting plants. Two main factors are contributing to this effect: the increase in atmospheric concentrations of carbon dioxide and the change in the composition and diversity of vegetation. The first factor stimulates soil microbes via increased plant photosynthesis and transfer of photosynthetic carbon to soil. Whereas, the second factor alters the amount and quality of organic matter entering soil, as well as other soil properties, affecting the activity and structure of belowground communities (Hartley et al. 2008).

On another hand, it is useful to indicate that the increasing of carbon sequestered in soil plays a key role in the change mitigation. Accordingly, several studies (e.g. Dijkstra et al. 2006; To et al. 2010; De Deyn et al. 2011) have identified suitable management practices of arable and degraded soils to increase carbon sequestration and the potential for forest soils to sequester carbon. In order to identify these practices, many studies have focused on the mechanisms involved in plant modulation of soil carbon sequestration that involve a myriad of biotic interactions between plants, their symbionts (i.e., mycorrhizal fungi and nitrogen-fixing bacteria), and decomposer organisms, whose activities determine the rate of decomposition of organic matter and hence carbon loss from soil (Orwin et al. 2010). Usually, the conceptual approach developed by De Deyn et al. (2011) is used to evaluate the relationship between plant communities, the microbial community, and soil carbon sequestration. This conceptual approach is consistent with the growing recognition that plant traits are major influences on soil nutrient and carbon cycling, and that certain plant traits can select for particular groups of soil organisms that play key roles in these mechanisms (Dijkstra et al. 2006).

2.5 CONCLUSIONS

In this chapter, we evaluated the interactions between soil and plant, and then we discussed how plant–climate–soil interactions can favor soil carbon sequestration. It is clear from the findings of this chapter that plant root carbon transfer to the soil and resulting carbon cascades through the plant–microbial–soil system play a primary role in driving carbon-cycle feedbacks and in regulating ecosystem responses to climate change. However, further research is required to understand the ecosystem function and response to climate change and the interactions between plant, climate, and soil processes. The aim of future work will be to identify suitable soil and plant management and climate mitigation strategies.

APPENDIX 2.1 (QUESTIONS WITH ANSWERS)

QUESTIONS

Question 1: What are the factors affecting the rate of diffusion?
Question 2: How do the relatively immobile nutrients ever make it to the plant roots?

ANSWERS

Answer 1: There are many factors affecting the rate of diffusion; the main factors are:

- Gradient of concentration: rate of diffusion increases with the increase in concentration gradient across the barrier. Diffusion stops when the concentration of the substances on either side of the barrier becomes equal;

- Permeability of membrane: rate of diffusion increases with the increased permeability of the membrane separating the two diffusing substances;
- Temperature: rate of diffusion increases with an increase in temperature;
- Pressure: this factor plays an important role in the diffusion of gases as gases diffuse from a region of higher partial pressure to a region of lower partial pressure.

Answer 2: The plants create a zone directly next to the root that has very low concentrations of these immobile nutrients. This allows diffusion to occur, which pulls nutrients that are further away from the root toward the root. This, in turn, pulls more of these immobile nutrients off the soil surface to maintain a balance between nutrients in solution and nutrients on the surface of the soil.

3 Soil Processes and Soil Properties

3.1 INTRODUCTION

Soil formation from a parent material (i.e. pedogenesis) is a complex process, which takes several million years. The formation of a particular type of soil depends on the properties of the parent material, weather, and other parameters (Brady and Weil 2002). Soil formation processes and the resulting soil properties can significantly affect crop growth (Carter 2002). Therefore, they need to be properly monitored to avoid destroying this essential building block of our environment and our food security. The formation and properties of soils are key areas of research in the soil science discipline. In this chapter, we first discuss the soil forming factors and processes. Then, we list and define the basic soil properties and understand the relationships between these properties.

3.2 SOIL FORMING FACTORS AND PROCESSES

3.2.1 Soil Forming Factors

In the 19th century, several soil scientists (e.g. Ruffin 1832; Lawes et al. 1883; Merrill 1906) considered that parent material (e.g. geology) was the main factor of soil formation. Currently, scientists (e.g. Blume 2002; Buol 2010; Soil Survey Staff 2010) consider that soils form over time under the influence of five main factors: parent material, time, climate, relief, and organisms (Figure 3.1). These five factors interact to form the soil profile:

- Parent material:
  The parent material refers to the mineral or organic material from which the soil is formed (i.e. origin of soil). It could be solid rock or deposits (e.g. alluvium and boulder clay). The soil that was weathered directly from the parent material is called residual soil. The parent material contributes significantly to the physicochemical properties of the soil, i.e. soils will carry several physicochemical properties of its parent material (e.g. color, texture, structure, etc.). For example, soils formed from granite are often sandy, whereas those formed from basalt are usually clay soils (Soil Survey Staff 2010). Furthermore, parent material also determines the rate at which soil forming processes occur (Bockheim and Gennadiyev 2000). Also, the

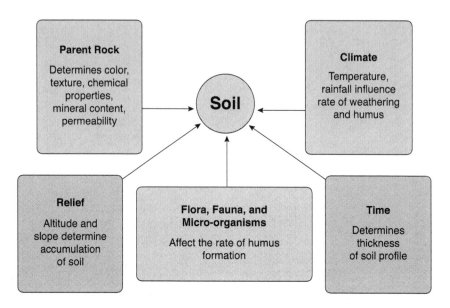

FIGURE 3.1 Soil formation factors.

minerals in the parent material play a key role in soil fertility and other soil properties.

- Time:

  The soils pass through a number of stages (steps) as they form, for this reason, time is considered as an important factor of soil formation (Buol et al. 1973). Over time, the soils may change from one soil type to another. These steps require a long period of time (several million years) and result in a deep soil profile with many well differentiated layers known as soil horizons (Figure 3.2). The topsoil (A horizon) is usually rich in dark colored organic remains called humus (labeled O horizon in Figure 3.2). The subsoil (B horizon) contains minerals that have been transported deeper by groundwater. Most of the clay in soil has also washed down to this layer. The partially weathered bedrock (C horizon) is composed of broken up bedrock on the top of bedrock (parent material) (Stockmann et al. 2011). The thickness of each soil horizon is variable, depending on the factors that influence soil formation.

- Climate:

  The climate variables (especially temperature, wind, and rainfall) are the most significant factors that affect soil formation (Scharpenseel et al. 1990). For example, seasonal patterns of heat flux and runoff can affect the depth and pattern of removal and accumulation of soluble and colloidal constituents in soil. Extreme rainfall can contribute

Soil Processes and Soil Properties

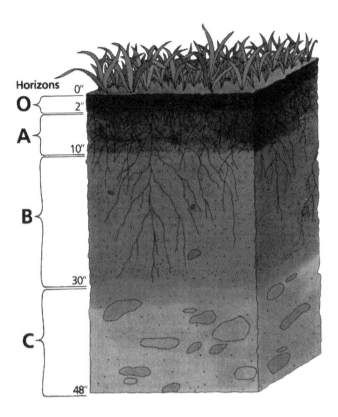

FIGURE 3.2 Typical horizons of soil profile.

to significant erosion of parent material. In Saharan regions, wind speed contributes to the redistribution of sand and other particles, often carrying them vast distances.

- Relief:

This factor refers to the position of a soil on landscape. The slope of the landscape has a direct contribution to the effects of climate factors (especially rainfall and wind) on the soil quality. Slope determines how soil is moved. Generally, landscapes with long slopes (uplands) have faster movement of water and have the potential to erode the soils highly. As a consequence, the effect will be soils with poor quality on the slopes and richer deposits (soils richer with mineral accumulation and organic matter) in the lowlands (Gillot 1981). Furthermore, soils in the lowlands will have more water than soils on the uplands. Accordingly, soils on slopes that face the sun's path will be drier than soils on slopes that do not (Rice 2003). This dry condition could have a negative impact on plant growth due to the close relationship between soil moisture conditions and plant growth (Greenland 1997).

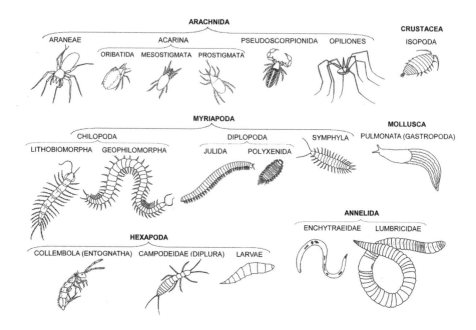

FIGURE 3.3 Common organisms that can live in and on the soil.

- Organisms:

  This factor refers to the crucial contribution of several organisms living in and on the soil to soil formation. The common organisms that can live in and on the soil are mentioned in Figure 3.3. These organisms serve to decompose plant materials and mix soil through the bioturbation process. The main result of this process is the formation of organic matter, which is considered as the richness of soil (Briones et al. 2010).

  Furthermore, the organisms contribute to soil aeration and help break down organic matter and aid decomposition. In addition, they help with mineral and nutrient cycling and chemical reactions (Greenland 1997). On another hand, soil biology affects mineral accumulation in the soil. Biologically mediated chemical weathering can create striking differences in color stratification (Soil Survey Staff 2010).

3.2.2 Soil Forming Processes

Several soil processes result from the interactions between the five soil forming factors discussed above (i.e. parent material, time, climate, relief, and organisms). These soil processes can be classified into four basic groups: addition, loss, transformation, and translocation. These soil forming processes, impacted by various environmental factors, give a logical framework for identifying the possible interactions between a particular soil and the landscape in which it functions.

- Addition:

  In this process, first water is added through rainfall, then, several new organic materials such as decomposed plants, organisms, and organic matter are added to the soil. Finally, some new mineral deposits (e.g. sand, silt, clay) can be added by some transport factors such as water and wind (Soil Survey Staff 2010);

- Loss:

  It refers to the removal of some elements from the soil profile. In this process, water is lost through evapotranspiration and mineral material through erosion. Also, organic material loss can occur when the erosion washes away topsoil and humus. Organic material may decompose into carbon dioxide;

- Transformation:

  The chemical weathering of sand and formation of clay minerals, transformation of coarse organic matter into decay resistant organic compounds (humus);

- Translocation:

  It refers to the movement of mineral and organic materials of soil between the various horizons of soil profile. Often, this movement is vertical.

3.3 SOIL PROPERTIES

Each soil is a mix of four components: mineral material, organic matter, water, and air (see details in Figure 3.4). The combinations of these components determine the properties of the soil (Balba 1995).

These soil properties are classified in the three basic groups: physical, chemical, and biological properties (Arshad et al. 1996). Soil fertility depends highly on these physical, chemical, and biological properties.

3.3.1 Physical Soil Properties

They mainly include the following properties: texture, structure, and porosity.

- Soil texture:

  It is the relative percentage of sand, silt, and clay particles that makes up the mineral material of the soil. The size (i.e. diameter) is the main criterion used to distinguish these particles (Table 3.1).

  Using "Textural triangle," and based on the relative percentage of sand, silt, and clay particles, several soil textural classes could be distinguished (Figure 3.5). Each soil textural class has its proper characteristics. For example, a soil which has a sandy textural class is easy to cultivate, has sufficient aeration for excellent root growth and is

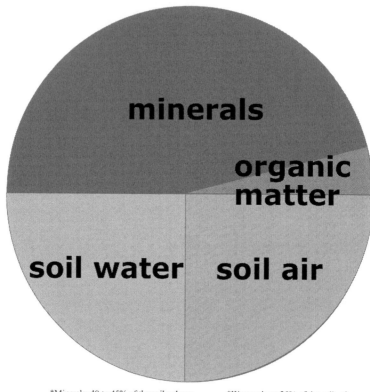

*Minerals: 40 to 45% of the soil volume   *Water: about 25% of the soil volume
*Organic matter: 1 to 5% of the soil volume   *Air: about 25% of the soil volume

FIGURE 3.4 Main soil components.

TABLE 3.1

Diameter (size) of Sand, Silt, and Clay Particles

Soil Particle	Diameter (mm)
Sand	0.05–2.0
Silt	0.002–0.05
Clay	<0.002

easily wetted, but it dries rapidly and easily loses crop nutrients through leaching.

Soil texture is the result of the physicochemical breakdown of the parent material under the influence of climate factors. Also, soil texture

Soil Processes and Soil Properties

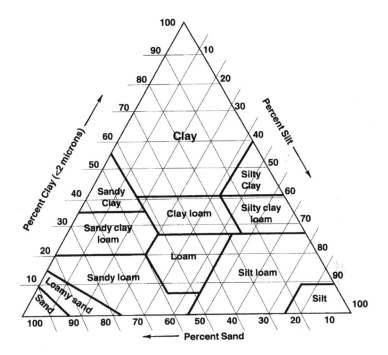

FIGURE 3.5 Textural triangle showing soil textural classes.

could be determined through the hydrometer method. It is a quantitative measurement providing estimates of the percent sand, clay, and silt in the investigated soil (see details in Gee and Bauder 1986).

- Soil structure:

  This property describes the arrangement of the sand, silt, and clay particles in the soil and of the pore spaces located between them (Gee and Bauder 1986). The soil particles are rearranged in aggregates (i.e. aggregation mechanism) based on several processes such as the precipitation of oxides, hydroxides, carbonates, and silicates; the products of biological activity (e.g. biofilms); ionic bridging between negatively charged particles (e.g. clay minerals and organic compounds) by multivalent cations; and the interactions between organic compounds. A good aggregation mechanism reflects a good soil structure and contributes to positive benefits for plant growth (White 1979). Therefore, improving soil structure could greatly help in improving plant growth.

- Soil porosity:

  Soil porosity refers to the percentage of the total volume of pores (micro and macropores) in a soil material. Accordingly, it is calculated as the ratio of the volume of pores to the total volume of the soil (Equation 3.1):

$$n = \frac{V_p}{V_s} \qquad (3.1)$$

where n is the soil porosity (–), V_p is the pores volume in the soil (m^3), and V_s is the total soil volume (m^3). When all pores are full of water, the soil is under saturation condition in which the water in the macropores will drain freely from the soil due to the gravity process. Several soil processes (e.g. aggregation mechanism) take place in soil pores. Accordingly, soil structure could have an impact on soil porosity by determining for example the size of the pores.

The three physical soil properties discussed above have direct impacts on water and air movement in the soil, with subsequent effects on plant water use and growth. Therefore, there is a need for continuous monitoring of these soil properties to ensure optimum plant growth in agricultural fields.

3.3.2 Chemical Soil Properties

These mainly include the following properties: micro and macronutrients, cation exchange capacity (CEC), soil pH, soil organic matter, and soil electrical conductivity.

- Micro and macronutrients:

    They refer to the essential mineral nutrients for plant growth. This is mainly assured by the macronutrients (calcium, magnesium, nitrogen, phosphorous, potassium, and sulfur) and also by the micronutrients (chlorine, boron, manganese, zinc, and iron) that are needed by the plants in much smaller amounts (Zhang et al. 2010). These essential nutrients are usually derived from the soil solution. Nutrient elements are taken up by plants in an "ionic" form (see Table 3.2).

    Soil fertility greatly depends on the availability of these ionic nutrients. Accordingly, in order to increase the soil fertility, some fertilizers are usually added to crops. An optimal combination between the available macronutrients and micronutrients in the soil and the amount of fertilizers to add facilitate nutrient uptake by plants, hence, contribute significantly to optimum plant growth (Yuan and Li 2007).

- Cation exchange capacity (CEC):

    CEC is the total capacity of a soil to hold exchangeable cation elements. The common exchangeable cations are calcium (Ca^{2+}), magnesium (Mg^{2+}), sodium (Na^+), and potassium (K^+). Cation exchange is a process in which a cation attached to a colloidal complex can exchange with the ions of the soil solution (surrounding liquid). This process is responsible for the availability of plant nutrients in the soil.

Soil Processes and Soil Properties

TABLE 3.2
Ionic Forms of Major Nutrient Elements

Element	Ionic Form
Nitrogen (N)	NO_3^-, NH_4^+
Potassium (K)	K^+
Phosphorous (P)	$H_2PO_4^-$, HPO_4^{-2}
Calcium (Ca)	Ca^{2+}
Magnesium (Mg)	Mg^{2+}
Sulfur (S)	SO_4^{2-}
Chlorine (Cl)	Cl^-
Iron (Fe)	Fe^{2+}, Fe^{3+}
Manganese (Mn)	Mn^{2+}
Boron (B)	$H_2BO_3^-$
Zinc (Zn)	Zn^{2+}
Copper (Cu)	Cu^{2+}

For this reason, CEC is considered as the best indicator of soil fertility (Lindsay 1979). Several factors could have impacts on CEC. The main factors are:

- Soil texture: for example, fine soil (e.g. clay) tends to have higher cation CEC than coarse soil (e.g. sand);
- Organic matter amount: an increase of organic matter amount in a soil favors higher CEC values;
- Soil pH: CEC increases with pH. Soil pH can affect CEC by altering the surface charge of colloids. Higher pH levels will neutralize the negative charge on colloids, thereby decreasing CEC. The opposite occurs when pH increases. Often, it is highly recommended for the CEC of a soil at a pH of 7.0.

CEC is calculated as the amount of exchangeable cations in 100 grams of soil (Equation 3.2) and expressed in millequivalents per 100 grams of soil (meq/100g). So, in order to calculate CEC, it is recommended to express the concentration of each exchangeable cation in millequivalents. In Table 3.3, we summarize the millequivalents (meq.) of common exchangeable cations and their equivalent in ppm. Low CEC values indicate that soils hold fewer nutrients, and will likely be subject to leaching of mobile "anion" nutrients. Based on the CEC values, several features of the agricultural soils and some practical applications can be determined. Furthermore, due to the importance of the CEC term, at the end of this chapter (in Appendix 3.1), an exercise (with correction) about how to calculate CEC is presented.

TABLE 3.3
Millequivalents of Common Exchangeable Cations and their Equivalent in ppm

Cation	Atomic Weight	Valence	Millequivalents	Equivalent (in ppm)
Ca^{2+}	40	2	20	200
Mg^{2+}	24	2	12	120
Na^+	23	1	23	230
K^+	39	1	39	390
NH_4^+	18	1	18	180
Al^{3+}	27	3	9	90
Zn^{2+}	65	2	32.5	325
Mn^{2+}	55	2	27.5	275

- Soil pH:

    It is the measure of hydrogen ions (H^+) in the soil. It reflects the acidity ($0 \leq pH \leq 7$) or alkalinity ($7 < pH \leq 14$) of soils. Usually, slightly acidic to neutral soil (pH 5.5–7) is a good condition for perfect plant growth. Silicate and carbonate minerals (that containing Ca^{++}, Mg^{++}, Na^+, and K^+ ions) contribute significantly to soil alkalinity increase. Soils become acidic when basic ions (especially Ca^{++}, Mg^{++}, Na^+, and K^+) held by soil colloids are replaced by hydrogen ions. Soil pH contributes directly and indirectly to the all soil chemistry reactions in the soil. For example, it contributes to the cations exchange and to nutrient availability in the soil, for this reason, the soil pH can have large effects on plant growth. Generally, a decrease of soil pH favors higher nutrient availability in the soil (van Breemen et al. 1983).

- Soil organic matter:

    It plays a key role in soil fertility. It is the fraction of the soil that consists of plant or animal tissue in various stages of decomposition. This latter depends highly on soil temperature, soil water content, aeration, pH, and nutrient levels. The final stage of the decomposition allows the formation of a stabilized organic matter called humus. Often, the organic matter occupies 1 to 6% of the total soil mass. Soil organic matter (especially humus) has several positive effects on soil such as increasing the cation exchange capacity of soil, improving water infiltration and soil aeration, improving soil ability to resist pH change, and providing food for the living organisms in the soil (Cotrufo et al. 2013). Due to these positive effects, several fertility amendments (e.g. animal manure and compost) are usually added to the agricultural soil to increase the organic matter level (Grandy and Neff 2008).

- Soil electrical conductivity (EC):

  Soil electrical conductivity (EC) is used to determine the amount of salts in an agricultural soil (i.e. soil salinity). It is an important indicator for soil health. EC is the ability of a soil to conduct an electrical current. The cause of the electrical conductance is the existence of particles with electric charges which are loosely bound to specific positions within materials and, thus, are capable of conveying electric charge. EC can be measured through various types of sensors based on different electromagnetic phenomena. EC is usually expressed in units of milliSiemens per meter (mS m^{-1}). Several cations and anions, especially Ca^{2+}, Mg^{2+}, Na^+, and K^+, Cl^-, SO_4^{2-}, HCO_3^-, CO_3^{2-}, and NO_3^-, contribute significantly to soil salinity increase. An intolerable level of EC can have serious negative impacts on crop growth (Ahmad and Sharma 2008). For example, it restrains water uptake from the roots by increasing the osmotic pressure of soil water, thus making it more difficult for plants to uptake nutrients. Also, some levels of EC may cause specific ion toxicities and imbalances in the nutritional stability of plants (Corwin et al. 2007).

3.3.3 Biological Soil Properties

The most common biological components of soil are earthworms, nematodes, microarthropods, bacteria, fungi, and actinomycetes. The main activities performed by these organisms are: mineralization, nitrification, nitrogen fixation, and denitrification (Ingham et al. 1985). All living organisms in the soil can significantly impact many soil properties. For example, they break down organic matter and while doing so make nutrients available for uptake by plants. Also, the nutrients stored in these organisms prevent nutrient loss by leaching. Furthermore, microbes maintain soil structure, and earthworms are important in bio-turbation in the soil (Baker et al. 2006).

As organisms are living in the soil, it is evident that some soil management practices can have negative impacts on biological properties in soils, some of the soil organisms being extremely sensitive to soil management. For example, soil microbial activity can be significantly increased by drainage practice. Hence, soil biological properties are considered as indirect indicators of suitable soil management and good soil quality, such as soil respiration rate or enzymatic activities that can be derived from living organisms in soil (Bardgett 2005).

3.4 CONCLUSIONS

Soil science is essential for all aspects of soil use and management. This chapter has presented an overview of the fundamentals of soil science. Although several useful data with respect to soil processes, soil horizons, and soil properties have been identified, with the continual changing of production practices

Climate Change Impacts on Coastal Soil and Water Management

and increasing needs from crop production (such as biomass and energy), continual monitoring of all soil properties and developing of suitable soil use and management are so needed to maintain sustainable agricultural production.

APPENDIX 3.1 (EXERCISE WITH CORRECTION)

EXERCISE:

On August 1, 2015, in a Tunisian agricultural field, a soil sample was collected to measure the cation exchange capacity (CEC). The laboratory measurement of cations concentrations had showed the following results: Ca^{++} (1 mg g^{-1}); Mg^{++} (0.135 mg g^{-1}); Na^+ (0.46 mg g^{-1}); and K^+ (1 mg g^{-1}).

1. Based on these results and using the data in Table 3.3, calculate the CEC of this soil.
2. Based on the calculated CEC value, identify the main feature of the investigated soil.

CORRECTION:

1. In a first step, we have to convert the cations concentrations expressed in mg g^{-1} to mg/100 g soil, then, to meq/100g (using the data presented in Table 3.2). So:

   *Ca^{++} (1 mg g^{-1} = 100 mg/100g soil = 5 meq/100 g)
   *Mg^{++} (0.24 mg g^{-1} = 124 mg/100 g = 2 meq/100 g)
   *Na^+ (0.46 mg g^{-1}= 46 mg/100 g soil = 2 meq/100 g)
   *K^+ (1.17 mg g^{-1} = 117 mg K/100g soil = 3 meq/100 g)
   *Mg^{++} (0.135 mg g^{-1} = 13.5 mg/100 g = 1.5 meq/100 g)

   In the second step, CEC value is found by summing up the cation concentrations (expressed in meq/100 g). So: CEC = 5 + 2 + 2 + 3 + 1.5 = 13.5 meq/100 g.

2. CEC = 13.5 meq/100 g; it is a good value (> 10) and reflects a good capacity of the soil to hold nutrient elements.

4 Aspects of Global Climate Change

4.1 INTRODUCTION

During the recent decades, the amounts of global greenhouse gases (i.e. carbon dioxide, methane, and nitrous oxide) in the atmosphere have greatly increased due to the intensification of industrial activities (IPCC 2013). This increase causes warming of the earth surface (i.e. temperature increase) and alters the energy transfer between atmosphere, space, land, and the oceans (Figure 4.1).

Since 1926, the global average air temperature has increased by about 0.9 °C and even more in some sensitive polar regions (IPCC 2013). Because the temperature of the earth surface is the driving force of earth's weather patterns, further change in other climate variables (such as rainfall, wind, air humidity, etc.) has also occurred. The change in all climate variables is referred to as global climate change (Rizzi et al. 2016). The signs (i.e. aspects) of global climate change are everywhere, and are more complex than just climbing temperatures. The major global climate change aspects include increased air temperature, changes in rainfall patterns, sea level rise, ocean acidification, floods, and droughts (Arnell et al. 2015). In the chapter, we discuss all these aspects.

4.2 ASPECTS OF GLOBAL CLIMATE CHANGE

4.2.1 INCREASING AIR TEMPERATURE

After a long tracking of the global average temperature (from 1901 to 2001), it was noted that the earth is warming (Figure 4.2).

As mentioned in Figure 4.2, the global average air temperature has increased by about 0.8 °C along the observed 100 years. The temperature increase began in 1910, and then highly increased in the 2000s (IPCC 2013). Based on modelling estimations (Figure 4.2), the temperature will continue to increase progressively to rise as much as 2 °C by the end of the 21st century (Clarke and Smethurst, 2010). The main cause of the global average air temperature is the increase in the amount of global greenhouse gases (i.e. carbon dioxide, methane, and nitrous oxide). There are several sectors that have resulted in increased greenhouse gases emissions. The following are the most common sectors (Table 4.1):

- Electricity sector:
  It is the largest contributor to global greenhouse gases emissions and accounts for 35% of total gases emissions (Bölük and Mert 2014).

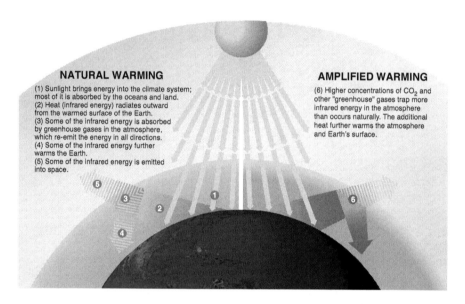

FIGURE 4.1 Warming of earth surface.

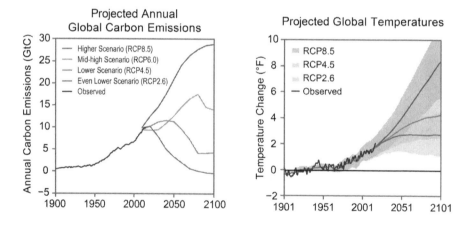

FIGURE 4.2 Observed and projected global average temperature from 1901 to 1921.

Carbon dioxide (CO_2) is the major emission of the electricity sector. This gas is mainly derived from fossil fuels used for electricity. Since 1990, the carbon dioxide emission from the electricity sector has increased by 12%. Furthermore, the global average annual concentration of CO_2 in the atmosphere has increased from 358 ppm in 1990 to 407 ppm in 2017 (IPCC 2013). A recent assessment of the effect of fossil

TABLE 4.1
Contribution of Various Sectors to Greenhouse Gases Emissions (from: IPCC 2013)

Sector	Contribution (%)
Electricity	35
Transport	29
Agriculture	27
Others	<10

fuel use on global temperature increase has showed that CO_2 emitted from coal combustion contributed to 0.3 °C of the 1 °C increase in global average annual surface temperatures above pre-industrial levels (Cole et al. 2005). Recently, some changes in the coal sector have been implemented to fight against global climate change. These measures include especially the improvement of energy efficiency in power plants by switching from coal to natural gas. Some power plants have begun using pulverized coal rather than conventional coal because it requires less coal to produce the same amount of energy. Improvement of energy efficiency and electricity generation from nuclear power stations could also be a powerful solution to reduce CO_2 emissions from the electricity sector (Lino and Ismail 2011).

- Transport sector:

  It is the second largest contributor to global greenhouse gases emissions and accounts for 29% of total emissions. In this sector, the greenhouse gases are mainly derived from fossil fuels used by transport vehicles such as planes, trains, automobiles, buses, trucks, and ships, and especially private vehicles. Carbon dioxide (CO_2) is the major emission of the transport sector (Zhou et al. 2013). The USA is the world's largest transportation energy consumer (IPCC 2013).

- Agriculture sector:

  It is the third largest contributor to global greenhouse gases emissions and accounts for 27% of total emissions. This total percentage is divided as follows: methane accounts for 50% of total agricultural emissions, 35% is from nitrous oxide, and 15% from carbon dioxide (IPCC 2013). The major sources of greenhouse gases emissions in the agriculture sector include enteric fermentation (i.e. microbial action in the digestive system), intensified energy utilization, and soil management (especially manure adding). There are many agricultural practices that can help decrease greenhouse gases emissions, including rotational grazing, reduced rates of fertilizers, good manure management (to decrease methane and nitrous oxide concentrations in the soil), and reduced energy utilization (e.g. using electrical motors instead of fossil-fuel motors) (Pervanchon et al. 2002; Forster-Carneiro et al. 2013).

4.2.2 Rainfall Patterns Change

As confirmed by several international studies (e.g. Ge et al. 2010; Sharad Jain and Kumar 2012; Al Charaabi and Al-Yahyai 2013), climate change is altering rainfall patterns worldwide. Increased air temperature has contributed to more moisture evaporation from land and water into the atmosphere (Figure 4.3). For each 1 °F increase in air temperature, the atmosphere can hold about 4% more water vapor. This evaporation process has in turn contributed to more rainfall and more heavy downpours. However, this extra rainfall is not spread evenly around the globe, and some places might actually get less rainfall, because climate change causes shifts in air and ocean currents, which can considerably change rainfall patterns.

Based on the past climate observations and future climate projections (through validated climate models), in general, wet areas get wetter, dry areas get drier. Accordingly, the effects of changing rainfall patterns on water resources will be

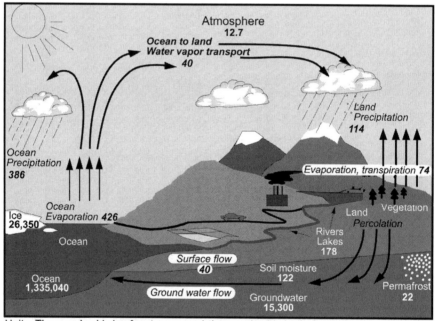

FIGURE 4.3 Estimates of the current global water budget and its annual flow using observations from 2002–2008 (adapted from Trenberth et al. 2011.)

Aspects of Global Climate Change   39

more serious in arid regions (Leung et al. 2004). In these regions, the implications of climate change on water resources can be one of the biggest risks to ensuring uninterrupted water supply for many purposes (such as domestic demand, agriculture, and industry).

4.2.3 Sea Level Rise

It is one of the most indirect effects of climate change in the coastal regions. After a long tracking of the global average sea level (from 1990 to 2010) through satellite radar measurements, it was observed that the world's sea level was rising highly (Figure 4.4).

As mentioned in Figure 4.4, the world's sea level rose by about 20 cm from 1990 to 2010. Accordingly, the rate of rise is about 1 cm year^{-1} (Figure 4.4) which is mainly due to increased global air temperature that drives the thermal expansion of seawater and the melting of land-based ice sheets and glaciers (Vermeer and Rahmstorf 2009). Furthermore, based on the future estimations of climate scientists, it is expected that the annual rate of sea level rise will continue to increase during the 21st century (Rahmstorf 2012).

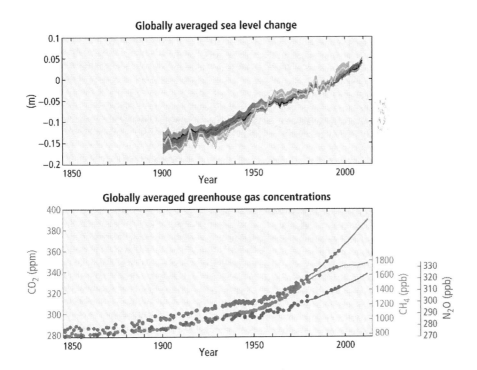

FIGURE 4.4 Global average sea level from 1990 to 2010 (adapted from IPCC 2013.)

Precise projections of sea level are difficult because of the complex properties of the climate system. However, past observations coupled with the present measurements of sea levels have improved computer climate models and sea level projections, up to 2100. The main aim of all developed models is to describe quantitatively, and with high accuracy, the various physical processes that lead to sea level rise (see details in Rahmstorf 2012).

There are several practical measures that can be practiced to fight against sea level rise in coastal regions. Three major measures can be distinguished: retreat measures, accommodation measures, and protection measures. In the retreat measures, further development of infrastructure in coastal areas at high risk must be avoided. Accommodation measures make societies more flexible to sea level rise; for example, the cultivation of crops that tolerate high soil salinization levels. With protection measures, coastal areas can be protected by the construction of dams, dikes, and by improving natural defenses (Thomsen et al. 2012).

4.2.4 Ocean Acidification

It refers to the progressive decrease in the pH of oceans over the recent decades (Jacobson 2005). The major source of this progressive acidification is uptake of atmospheric carbon dioxide (CO_2) in the oceans (details in Figure 4.5).

Ocean water is slightly basic (7.1 to 7.3), and with progressive acidification, the basic condition can convert to acidic condition (pH < 6.9). This new condition is expect to have a range of potentially harmful effects for marine organisms, such as depressing metabolic rates, reducing nutrient availability,

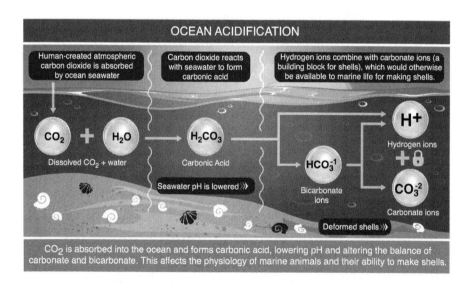

FIGURE 4.5 Ocean acidification process.

and immune responses in some organisms (Feely et al. 2004). Acidification has resulted in a significant modification of the saturation state of the oceans with respect to calcium carbonate ($CaCO_3$) ions, causing carbonate ions to be relatively less abundant in the oceans (Caldeira and Wickett 2003). Hence, estimating $CaCO_3$ dissolution rates for the global oceans from total alkalinity and chlorofluorocarbon data is essential for monitoring ocean acidification.

The ocean absorbs more than 50% of the CO_2 that is available in the atmosphere, and as atmospheric CO_2 concentration increases there will be a significant increase in the oceanic CO_2 concentration (IPCC 2013). Due to the emissions of the electricity, transport, and agriculture sectors (details in section 4.2.1), the CO_2 concentration in the atmosphere has reached very critical values. In 2018, this concentration reached about 411 parts per million (ppm) per month (IPCC 2013). Therefore, it is necessary to move to new principles of combustion, to new technologies where carbon dioxide and natural water will act as a fuel-oxidizing pair.

4.2.5 Floods

Usually, the high risk of damaging flood events (Figure 4.6) is the result of the increase in the likelihood of extreme rainfall (Thodsen 2007). This has

FIGURE 4.6 Storm damage in Central Texas.

great impacts on life, existing infrastructure, and economic resources in the coastal regions, especially in the areas where the existing infrastructure has not been designed to cope with these threats (Doocy et al. 2013). For example, in the last decade, severe floods increased worldwide. These critical flood events reduce agricultural productivity in many agricultural coastal regions through significant degradation of soil characteristics (Gray et al. 2015). Therefore, greater attention to climate change is essential to assure sustainable production of food crops.

Due to the negative effects of damaging floods on coastal infrastructure, development of some adaptation measures is needed, especially in the areas at high risk of flood events. These measures must include practical actions to adjust storm water management infrastructure such as low impact development methods to reduce surface runoff or constructed storage to handle the increased flows during an extreme flood event. These adaptation measures are typically developed based on flood forecasting and hydrologic and hydraulic modeling (Thodsen 2007). The effectiveness of each practical measure depends on the accuracy of simulating projected impacts, such as the effectiveness of a flood control structure to protect a city from future increased flooding.

4.2.6 Droughts

Drought is a complex phenomenon, affected by changes in the hydrological cycle and resulting in various effects across many sectors, which leads to land degradation and forest dieback. In Figure 4.7, a schematic diagram showing drought propagation under climate change is presented.

Drought has various negative effects on hydroclimatic variables such as rainfall decrease, evapotranspiration increase, and soil moisture reduction (Mukherjee et al. 2018). Drought propagation has occurred due to climate change, especially due to the variability in anomalies to sea-surface temperatures. The effects of drought under climate change are likely to accelerate in the future (Salvadori and De Michele 2004).

Emissions of greenhouse gases have intensified droughts worldwide; however, the major intensification has occurred in arid regions. In these regions, evapotranspiration may produce periods of drought, defined as below-ormal levels of rivers, lakes, and groundwater, and lack of enough soil water content in agricultural fields. Based on future estimations with climate models, it is expected that the amount of land affected by drought will grow by mid-century and water resources in affected areas will decline as much as 30% (IPCC 2013). In arid and semi-arid regions, dry conditions quickly lead to water scarcity, emphasizing that the local climatology is also a factor. Hence, in water scarce or arid regions, where drought and water scarcity usually occur simultaneously, drought situations are more severe and further aggravate water scarcity (Van Loon and Van Lanen 2013). Accordingly, in some arid and semi-arid regions, the selection of suitable indicators that make a clear distinction between drought and water scarcity is necessary to make effective water management strategies.

Aspects of Global Climate Change

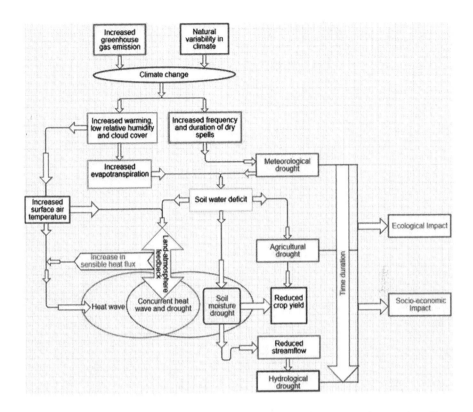

FIGURE 4.7 A schematic diagram showing the drought propagation under climate change (adapted from Mukherjee et al. 2018.)

4.3 CONCLUSIONS

In this chapter, we examined the basic aspects of climate change. Additional further work is needed to understand clearly the relationship between climate variables such as air temperature and rainfall and the contributing factors that govern these variables. In addition, further work is required to understand the impacts of climate change aspects on soil and water resources to develop suitable soil and water management strategies.

5 Impacts of Climate Change on Soil Resources

5.1 INTRODUCTION

According to the reports of the Intergovernmental Panel on Climate Change (e.g. IPCC 2007, 2013), climate change, especially increased air temperature and changing rainfall patterns, will certainly have a great impact on soil, thereby affecting crop growth and food security since soil has a key role in supplying macro and micronutrients to all crops (Brevik 2012). Furthermore, soil is considered as the most sensitive element to climate change due to its high vulnerability to small changes in climate variables, especially air temperature, rainfall, and evapotranspiration (IPCC 2013). Climate change is expected to affect soil resources by degrading many soil properties and through various land degradation processes (Karmakar et al. 2016). Understanding these effects is critical for developing ways to adapt to climate change. Therefore, in this chapter we will focus on how climate change may affect soil properties and land degradation processes.

5.2 EFFECTS OF CLIMATE CHANGE ON SOIL PROPERTIES

5.2.1 Effects on Physical Soil Properties

Climate change is likely to have effects on the following physical soil properties: soil texture, soil structure, soil moisture, and soil temperature (Karmakar et al. 2016; Gelybó et al. 2018). The hydrological aspect of soil such as water retention and hydraulic conductivity are significantly impacted by these physical soil properties, all of which contribute to the water, air, and heat management of the soil profile (Hillel 1973). The physical soil properties have significant effects on soil chemical and biological processes, such as adsorption, water transport, and nutrient uptake that play key roles in soil fertility (Horel et al. 2015). The effects of climate change on physical soil properties can include the following:

- Effects on soil texture:
  Soil texture refers to the relative percentage of sand, silt, and clay particles that make up the mineral material of the soil, and it has a direct effect of climate change. In Figure 5.1, we show the effects of climate scenarios on texture differentiation of soils. Soils with different texture

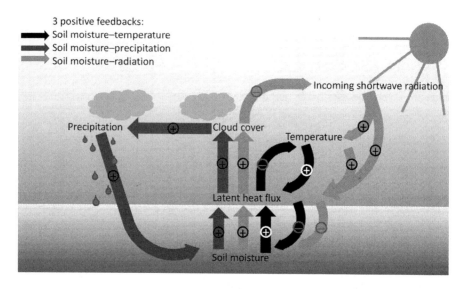

FIGURE 5.1 Soil moisture–atmosphere feedbacks. The plus and minus indicate positive and negative feedback loops, respectively. Soil moisture–temperature (black), soil moisture–precipitation (dark grey) and soil moisture–radiation (light grey) feedback loops.

had different reactions to changing climatic conditions, especially in the case of very critical climate conditions.

As confirmed by Gelybó et al. (2018), soils with clay texture are the most sensitive soils to climate change. Generally, climate change has various effects on cycles of wetting and drying in clay soils worldwide (Clarke and Smethurst 2010). Increased numbers of cycles of soil wetting and drying contributes highly to the cracking and shrinking mechanisms of clay soils (Brinkman and Brammer 1990). Indeed, repeated wetting and drying in clay soils highly contributes to the formation of cracks in the soil (Gelybó et al. 2018). Several studies (e.g. Varallyay 2007) have revealed that significant decreases in annual rainfall coupled with high increases in air temperature are likely to become the main condition in the 21st century in many regions of the world (Clarke and Smethurst 2010). This will increase the magnitude of the cycles of winter soil wetting and summer drying. The cracking and shrinking mechanisms occur usually in clay soils; however, water and soil losses due to rapid and non-uniform downward water movement through long conductive pathways could increase the number of drought events and intensive rainfall events. The cracking and shrinking mechanisms can have a very big *impact* on *soils* and the functions that *soil* performs. Therefore, careful monitoring of clay soils is required to avoid any further deterioration of soil properties (Clarke and Smethurst 2010).

- Effects on soil structure:

  It is an important soil property which describes the arrangement of the sand, silt, and clay particles in the soil and of the pore spaces located between them (Gee and Bauder 1986). Soil structure has direct and indirect effects of climate change. Water movement, nutrient uptake, soil fauna, and other processes depend highly on this soil property. The significant direct effects of climate on the soil structure are the destructive potential of raindrops, surface runoff and filtrating water, and extreme rain events (Varallyay 2010). Indirect effects are related to changes in the vegetation pattern and functioning, in soil use (Singh et al. 2011) and in the biological functioning of the soil (due to the sensitivity of earthworms, termites, and soil microorganisms to climate change), which also can have a negative effect on the amount and quality of soil organic matter. The soil structure is strongly affected by the amount and quality of soil organic matter, the mineral constituents of the soil, and the soil management practices (e.g. manure adding). A decrease in the amount of soil organic matter contributes to a decrease in soil aggregate stability and infiltration rates, and an increase in susceptibility to compaction, and furthermore susceptibility to erosion (Karmakar et al. 2016). Soil aggregate stability, soil porosity, and pore size distribution are usually used to define the structure of each type of agricultural soil. These parameters play a key role in the moisture and aeration status of the soil. Any changes in soil porosity can have direct effect on the soil water storage capacity (e.g. infiltration, water retention) and the aerobic (CO_2 emission) and anaerobic (CH_4 emission) conditions of the soil (Toth et al. 2018). Intensive rainfall (e.g. flood), and raindrops directly have a significant effect on soil aggregate stability (Varallyay 2007). Increasing temperature, coupled with lower available water, less biomass and soil organic matter amount can lead to an important reduction in aggregate size and stability (Karmakar et al. 2016).

- Effects on soil moisture:

  Soil moisture changes with climate change through a number of climate variables such as precipitation, which can increase soil water content within a few hours, and increased temperature which leads to greater evapotranspiration loss of water from the soil profile. The soil moisture regime reflects temporal changes in soil water content, the amount of soil moisture available for plants, and water loss through the leaching process and weathering, and is therefore a useful indicator of the soil forming environment (Karmakar et al. 2016). Several studies have evaluated the soil moisture regime under climate change (e.g. Hernádi et al. 2009; Holsten et al. 2009). These modeling studies used daily climate data of critical climate scenarios, especially A2 scenario (temperature increase and rainfall decrease) and B2 scenario (more frequent extreme climate conditions) and climate data for a reference past period. These studies have revealed that under critical climate scenarios a significant decrease (> 10%) in the average available soil moisture could be expected by the middle of the 21st century. Furthermore, the results of these studies showed that, compared to the present climate conditions, by the end of the 21st century a lower amount of water will be stored in agricultural soils worldwide, and this shortage of water will be as severe as in extremely dry years in the present climate. Therefore, there is a great need to develop various soil and water management options that could decrease the drought sensitivity of agricultural soils worldwide. Further understanding the effects of climate change on soil moisture is essential for the development of suitable soil and water management options (Gelybó et al. 2018).

- Effects on soil temperature:

  It is widely accepted that our earth has warmed as a direct result of climate change (IPCC 2013). Climate warming has various impacts on ecosystems and natural resources. Soil temperature is considered as the more sensitive element to climate warming. The soil temperature is a good indicator of the effects of climate change on soil properties (Karmakar et al. 2016) because of the close relationship between air temperature and soil temperature. Generally, increased air temperature causes an inevitable increase in soil temperature. Soil temperature is affected by several factors, including radiation at the soil surface, soil evaporation, heat conduction through the soil profile, and convective transfer (gas and water movements) (Wang et al. 2014). Interactions between soil temperatures and climate change have received less attention in climate effect studies, but a few studies have been published (e.g. Yang et al. 2010; Qian et al. 2011; Wang et al. 2014). These numerical simulation studies confirm the direct impact of the warming trend of air temperatures on increased soil temperatures. This warming trend of soil is expected to continue in the future (IPCC 2013). In addition, these studies have revealed that soil temperature increase could have several positive impacts on soil processes such as acceleration of organic matter decomposition, enhancement of microbiological activities in the soil, and increase of nitrification rate (Qian et al. 2011). However, soil temperature increase may also have negative impacts on soils depending on the changes in the timing and duration of the air temperature increase and on the type of vegetation occurring at soil surface. Therefore, there is an urgent need for proper monitoring of soil temperature regimes in relation to changes in air temperature and vegetation type (Zhang et al. 2005).

5.2.2 Effects on Chemical Soil Properties

Climate change is likely to have effects on the following chemical soil properties: soil pH, soil organic matter, and nutrient content (Karmakar et al. 2016; Gelybó et al. 2018):

- Effects on soil pH:

  Soil pH depends on many factors. The climate variables such as rainfall and temperature are among the most significant factors. Therefore, it is expected to change due to the effects of climate change (Gelybó et al. 2018). For example, increased rainfall can contribute to significant soil acidification which accelerates mobilization of toxic ions elements such as heavy metals, and can contribute to the depletion of basic cations through leaching (Brinkman and Brammer 1990) in soil. Furthermore, in a wetter climate, soil acidification can increase if buffering pools become depleted. On the other hand, temperature increase and low rainfall (i.e. arid climate conditions) may lead to capillary water movement and the evaporation of shallow groundwater, resulting in critical soil salinization levels (Varallyay 2007). There are several agricultural options that can be practiced to reduce acidification in the soil. The major agricultural options include the following management practices:

  - Reduce fertilizer application; especially, apply only the adequate fertilizer amount;
  - Decrease fertilizer amount applied in a single dose;
  - Use less acidifying nitrogen sources, e.g. calcium nitrate;
  - Increase irrigation efficiency to avoid leaching and increase nutrient retention.

- Effects on soil organic matter:

  The soil organic matter is the most important component of soil due to its positive effects on many soil properties such as structure formation, water-holding capacity, cation exchange capacity, and for the nutrients supply (Brevik 2009, 2013). The productivity and fertility of soils depends highly on soil organic matter. In the last years, climate change has already led to a significant decline in organic matter amounts worldwide which increases the susceptibility to soil erosion (Bot and Benites 2005). Carbon (C) and nitrogen (N) are the main components of soil organic matter. Several studies (e.g. Zaehle et al. 2010; Wan et al. 2011) have been interested to find responses to the following question: how potential changes in the C and N cycles (due to climate change effects) will influence soils? These studies concluded that temperature increase is likely to have a negative impact on C allocation to the soil, leading to an important decrease in soil organic C and contributing to a positive-feedback in the global C cycle. Furthermore, these studies showed that CO_2 increase contributes to enhancing the soil C:N ratio, because decomposing organisms in the soil need more N, which can decrease N mineralization (Gill et al. 2002; Reich et al. 2006). Mineralization is an important process in supplying N to crops (Pierzynski et al. 2009). Accordingly, if N mineralization is decreased, crop-available N levels in the soil and crop productivity would also decrease. The main recommendations of these studies include especially the need of further knowledge of the impacts of climate change on the N cycle, which has received far less attention than the C cycle. Currently, little is known about how climate change will affect soil organisms (IPCC 2013) which are essential in driving portions of the C and N cycles that take place in the soil.

- Effects on nutrient content:

  Several effects can occur depending on climate conditions. For example, drought conditions can contribute to significantly reduced nutrient acquisition capacity of root plants. Also, intense rainfall events in some regions can cause significant source of soil nutrient loss in the soil (Sun et al. 2007) such as nitrate leaching. Intense rainfall events, coupled with poor drainage practices in agricultural fields, can contribute to soil waterlogging and hypoxia which then contributes to nutrient deficiency and nitrogen losses (Drew 1988). On the other hand, soil temperature increase can enhance nutrient uptake by enlarging the root surface area and increasing rates of nutrient diffusion and water influx (Atwell and Steer 1990). However, the positive effects of soil temperature increase on nutrient content depend on adequate soil moisture. For example, under dry condition, higher temperatures result in extreme vapor pressure deficits that trigger stomatal closure (reducing the water diffusion pathway in leaves), and then nutrient acquisition, driven by mass flow, will decrease (Cramer et al. 2009). Emerging evidence suggests that soil temperature increase greatly affects nutrient status by altering plant phenology (Nord and Lynch 2009). Furthermore, soil warming accelerates loss of soil organic carbon.

5.2.3 Effects on Biological Soil Properties

The biological properties of soil are mainly related to the activities of thousands of living organisms in the soil (such as earthworms, nematodes, bacteria, and actinomycetes). These organisms are essential components of soil, and they play a key role in the retention, breakdown, and incorporation of plant remains, nutrient cycling, soil structure, and porosity. Increased air temperature may not have a direct impact on the ecological composition because these organisms have the ability to adapt with changes in air and soil temperatures. However, changes in ecosystems and migration of vegetation zones may seriously affect less migratory soil organisms through increased temperature and rainfall patterns change. In addition, the increase in atmospheric CO_2 levels in the soil typically facilitates the various activities assured by the soil organisms. These facilities include acceleration of nitrogen fixation, nitrogen immobilization, and denitrification, increase of mycorrhizal associations, increase of soil aggregation, and also increase of minerals weathering. These positive impacts enhance plant growth (Jones et al. 2009), but positive impacts depend on the balance between increased plant growth and increased decomposition of soil organic matter which will emerge under a changing climate.

5.3 CLIMATE CHANGE AND LAND DEGRADATION PROCESSES

Soil degradation induced by climate change refers to a change in soil properties caused by various climate variables. The climate change could have significant impacts on the following major land degradation processes: soil erosion, soil salinization, and soil acidification (Karmakar et al. 2016).

5.3.1 Soil Erosion Processes

The impacts of climate change on soil erosion process are very difficult to evaluate due to the contribution of several factors to this process (Gelybó et al. 2018). However, it is certain that this process, which is likely to be exacerbated due to climate change, leads to various negative consequences such as the displacement of soil (Figure 5.2), the displacement of soil organic carbon, and increasing carbon emissions to the atmosphere (IPCC 2013).

The rate, type and extent of soil erosion depend on several factors. The major factors include climate variables (especially rainfall and wind), vegetation characteristics (type, continuity, and density), and relief properties:

FIGURE 5.2 Displacement of soil.

- Erosion and climate variables:

  One of the most significant direct effects of soil erosion induced by rainfall is the aggregate destructing role of raindrops, surface runoff, and filtrating water. Decreased rainfall can reduce water erosion, whereas, the cohesion between soil particles can decrease due to moisture decrease which may contribute to accelerating soil erosion by wind factor. The latter may decrease significantly soil nutrient reserves by removing the finest soil particles, which contain the largest amounts of plant nutrients. Several studies have revealed the contribution of intensive wind erosion to significant decrease in soil organic carbon. Some critical organic carbon levels in the soil can contribute directly to important decrease in plant growth.

- Erosion and vegetation characteristics:

  Usually, intense rainfall leads to an important change in the extent of vegetation cover. Some agricultural practices such as irrigation and land use also lead to changes in the vegetation cover. For example, overgrazing, irrational land use, misguided agricultural utilization (cropping pattern, crop rotation), and inappropriate irrigation practices have negative impacts on vegetation cover. However, rational land use, appropriate irrigation practices may help to maintain or restore good vegetation cover.

- Erosion and relief properties:

  Relief properties (especially slope evolution) have direct effects on the velocity at which surface runoff will flow. Generally, steeper slopes (especially those without good vegetation cover) are more susceptible to erosion process during rainfall events than less steep slopes. Furthermore, steeper slopes are more prone to mudslides, landslides, and other forms of gravitational erosion (Egholm et al. 2009).

Simulations of the effects of climate change on soil erosion are subject to many uncertainties due to several factors including especially data availability and process knowledge. However, based on the performed simulation works, future climate change will increase soil erosion worldwide. General Circulation Models (GCMs) are common tools for climate change projection. Model simulations based on downscaled climate change projections (using GCMs with various global greenhouse gases emission scenarios) suggest a decrease in erosion, while large changes in daily annual rainfall amount usually resulted in a great increase in erosion.

5.3.2 Soil Salinization

In regions where the soil–water budget was previously balanced, intensifying evaporation processes may make this budget evaporation-dominant, leading to more intensive upward water movement and a high risk of salinization, especially in areas where saline shallow groundwater is close to the soil surface. This impact of climate change on soil salinization is more pronounced in coastal areas. As sea levels rise, low-lying coastal areas are increasingly inundated with saltwater, gradually contaminating the soil. These salts can be dissipated by rainfall, but climate change is also increasing the frequency of extreme climate events such as droughts and heat waves. This leads to more intensive use of costal aquifers for drinking and irrigation purposes, which further depletes the costal aquifer, causing more soil salinization. Some critical levels of soil salinization in agricultural areas affect plant yield in various ways. For example, it restrains water uptake from the roots by increasing the osmotic pressure of soil water, thus making it difficult for plants to extract water from the soil. Also, some soil salinization levels may cause specific ion toxicities and imbalances in the nutritional stability of plants. Depending on the soil salinization level, rice farmers can expect to lose more than 50% of their crop. For example, due to intolerable soil salinization levels, 13 million hectares of rain-fed rice were lost in South and Southeast Asia. The global distribution of salt-affected soil is presented in Figure 5.3. As shown in this figure, areas of Australia and Asia are highly affected by the soil salinization problem. In these areas, the extent of saline soils is 444 million hectares.

5.3.3 SOIL ACIDIFICATION

This process mainly occurs as a result of nitrate leaching in the rainfall regions (Fenton and Helyar 2007). The areas that show this process usually

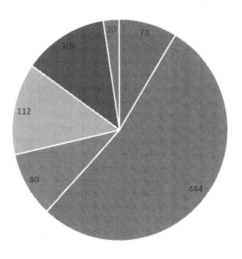

FIGURE 5.3 Global distribution of salt-affected soil (adapted from Shahid et al. 2018.)

have a low agricultural productivity due to the negative effects of acidified soils on plant growth. For example, acidified soils (especially when soil pH < 6) affect the availability of plant nutrients. Indeed, nutrients are often available to plants in the optimum 6 to 7.0 range. In several cases, unsuitable soil management practices in the rainfall areas aggravate the soil acidification (Charman and Murphy 2007). The extension of acidified soils depends on rainfall and temperature. Substantial increase in rainfall increases leaching and causes acidification, whereas a decrease in rainfall reduces intensity and extension of acidified soils (Wild 1993). Semi-arid and arid areas are influenced by seasonal changes, i.e. from leaching conditions to evaporative conditions. Acid sulfate soils are examples of acidification as they suffer from extreme acidity as a consequence of the oxidation of pyrite when pyrite-rich parent materials are drained (Rounsevell and Loveland 1992). Finally, improved understanding of the impacts of climate change on soil pH and other soil properties is required to develop a suitable soil management strategy to prevent loss of environmental function in soils and terrestrial ecosystems.

5.4 CONCLUSIONS

The state of any ecosystem is largely determined by the soil properties that are dependent on the climate in general. In addition, investigations on the land degradation processes can provide indicative results for many aspects of ecosystem functioning. Hence, we need to focus on exploring the expected impacts of climate change on the soil properties, and on the land degradation processes. In this context, modeling of climate change effects on soil properties and improved understanding of the impacts of climate change on land degradation processes could be perfect tools for developing a suitable soil management strategy to prevent loss of environmental function in soils and terrestrial ecosystems.

6 Impacts of Climate Change on Water Resources

6.1 INTRODUCTION

The increase in global air temperature coupled with changes in rainfall patterns will involve greater climate extremes, including higher intensity rainfall events or decreased streamflow conditions. Consequently, climate change could result in significant impacts on water resources, i.e. surface and groundwater systems (Schnorbus et al. 2014). Climate change impacts on these systems are the consequences of the influences of natural processes and human activities on the hydrologic cycle (Figure 6.1) (Dragoni and Sukhija 2008).

The hydrologic cycle is linked with changes in atmospheric temperature and radiation balance. Global air temperature increases are coupled with important changes in components of the hydrologic cycle such as changes in evapotranspiration, changes in atmospheric water vapor, and changes in rainfall patterns (Bates et al. 2008). Critical changes will seriously affect water resource availability, influence groundwater recharge and storage, and cause changes in surface water and groundwater quality (Green et al. 2007). Understanding these changes is essential for identifying ways to adapt to degradation of water resources. Therefore, in this chapter we will focus on how climate change may affect surface and groundwater systems.

6.2 EFFECTS OF CLIMATE CHANGE ON WATER RESOURCES AVAILABILITY

Climate change may have serious effects on water resources availability and management at the basin scale. These effects are more pronounced in the regions already suffering from water resources stress, such as the semi-arid and arid regions (Brekke et al. 2009). In these regions, the effects of climate change on the hydrologic cycle at the basin scale are main contributors for the significant decrease in water resources availability. Furthermore, in the semi-arid regions, several international studies (e.g. Lenderink et al. 2007; Senatore et al. 2011; Seiller and Anctil 2014) project a significant depletion of the water resources with a significant annual mean decrease in rivers discharge at the end of the present century. For example, Ducharne et al. (2010) have revealed that climate change will decrease the discharge of the Rhine River by 30% at the end of the century. On the other hand, several intense rain events (> 200 mm/hours) have

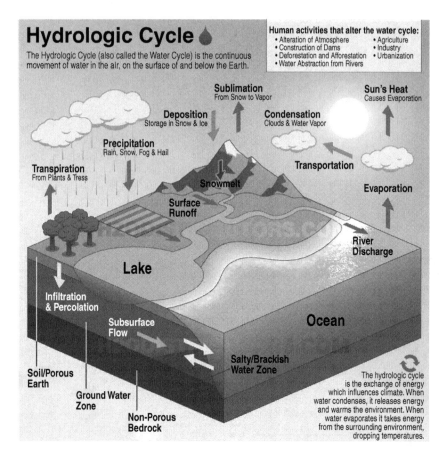

FIGURE 6.1 Influences of natural processes and human activities on hydrologic cycle.

been recorded in many humid regions (Younis et al. 2008). Because of these extreme events, water is relatively scarce throughout most of the year (Viviroli and Weingartner 2004). The irregularity in spatial distribution of rainfall contributes to significant disparities in water resources availability between various areas and makes water management more complex at the basin scale.

Under some circumstances such as the absence of shallow or deep groundwaters, it is essential to find suitable ways to store the available water from intense rain events to increase the availability of water resources throughout the year. The development of infrastructures such as the dams could also be helpful in this context. In recent years, the instability of water resources throughout time due to climate change effects obligated decision makers to develop large infrastructures (Figure 6.2) to resist intensification of rainfall events and increase water availability, especially for agricultural purposes.

Impacts of Climate Change on Water Resources   57

FIGURE 6.2   One of Japan's biggest dams.
Source: wikimedia.org

The sustainability of each ecosystem requires a good evaluation of the current and future availability of the water resources, especially at the basin scale. Key questions for managing well the available water resources are how much water is available, where is it available, and when? Also, the assessment of the spatial and temporal distribution of water resources and the effects of projected climate change on water resource availability at regional, national, or global levels could be so helpful to achieve the sustainability of water resources worldwide.

6.3   EFFECTS OF CLIMATE CHANGE ON SURFACE WATER QUALITY

Projected changes in temperature and rainfall as a result of the high concentrations of greenhouse gases in the atmosphere will directly and indirectly affect the physico-chemical quality of surface water. Indirectly, physical and chemical quality of water is expected to change due to the following processes:

bio-chemical reactions; nitrification and denitrification; acidification; salinization; and change in oxygen concentrations (Admiraal and Botermans 1989).

- Bio-chemical reactions:

  Increased air temperature typically leads to a significant acceleration of several chemical reactions. Accordingly, higher air temperatures which lead to warmer surface water, directly lead to an important increase in the rates of (bio) chemical processes. Humidity and sediment temperatures in the surface water (e.g. in a river) have a close relationship with microbial activity. Higher sediment temperature is expected to cause higher microbial activity (Van Dijk et al. 2009), which in turn will increase sediment respiration of organic material and then, concentrations of dissolved organic carbon in soils. Furthermore, the potential increase in rainfall amount will wash away the dissolved organic carbon from soils to surface waters. The resulting increase in dissolved organic carbon in surface waters will lead to a higher capacity to absorb light and hence higher temperatures. As a major result, when air temperatures increase, surface waters with high levels of dissolved organic carbon will face a higher increase in temperature than surface waters with low levels of dissolved organic carbon (Winterdahl et al. 2016).

- Nitrification and denitrification:

  Nitrogen plays a key role in each ecosystem. Nutrient productivity depends highly on this element. The presence of nitrogen in the surface waters with critical doses may cause eutrophic conditions, which adversely affect the quality of surface water. The processing of nitrate in surface water occurs by two major processes: nitrification and denitrification. Nitrification is the biological oxidation of ammonium (NH_4^+) to nitrate (NO_3^-) which is used by primary producers and is consequently considered a "nitrate" input process for surface water. Denitrification refers to the microbial process where nitrate (NO_3^-) is reduced and transformed to nitrogen (N_2) or nitrous oxide (N_2O) through a series of intermediate gaseous nitrogen oxide products. N_2 is released from the surface water into the atmosphere, resulting in a loss of nitrate from aquatic ecosystems. N_2O is a major greenhouse gas. The nitrification and denitrification processes either contribute to or remove nitrate from the surface water, they have significant effects on the physico-chemical quality of surface water. The rates of these processes are influenced by several factors such as organic matter concentration (Chen et al. 2018). Usually, an increase in organic matter concentration in the water (e.g., because of erosion during heavy rainfall) leads to significant distribution of the balance between nitrification and denitrification. The living microorganisms in the water mineralize the organic matter and nitrate concentrations increase.

- Acidification:

  It is well accepted that increased CO_2 concentrations in the atmosphere cause acidification of the surface waters, especially oceans (IPCC 2013).

Contrarily to ocean waters, several freshwater systems such as lakes receive more amounts of dissolved organic carbon from terrestrial ecosystems (Van Dijk et al. 2009). Bacterial activity mineralizes the dissolved organic carbon into CO_2. Therefore, O_2 concentrations in freshwater systems are often not in equilibrium with the atmosphere but are related to the concentration of dissolved organic carbon. Most freshwater systems are concentrated with CO_2. Consequently, an increase in the CO_2 concentration in the atmosphere is not expected to have an important impact on freshwater acidification (i.e. change in water pH). On the other hand, an increase in the temperature will promote more algal blooms. As a result, an increase of CO_2 from the water uptake occurs, and consequently, there is a significant increase in OH^- ions due to (bi-) carbonate equilibrium in the water. This condition makes the pH of freshwater systems more alkaline rather than more acidic (Van Dijk et al. 2009). In addition, pH increase due to increased soil erosion (which promotes more deposition of cations in the surface water) makes freshwater systems more alkaline. Under this situation, there is a positive effect of climate change on reduction of the acidification process in surface water (Parry 2000). The acidification of freshwaters results especially from acid rainfall. The emission of SO_2, NO_x, and NH_3 elements from the traffic, agriculture, and industry sectors is the main source of this acid rainfall.

- Salinization:

  Droughts and sea level rising are considered as the main aspects of climate change. These two consequences of climate change have highly contributed to the salinization of surface waters. Extreme drought events, which are becoming much more frequent and long-lasting due to climate change, lead to more pumping of water from surface water resources such as rivers for irrigation and water supply purposes. This excessive pumping can lead to increased salinization of surface water as salt gets concentrated. Coastal water resources (such as rivers connected to the sea and the coastal aquifers) are the most vulnerable to salinization due to the process of seawater intrusion. Sea level rise has significant effects on the salinization of these rivers and aquifers in low lying areas. In hot summers, river discharges will be low and the predicted sea level rise will increase the intrusion of seawater. Currently, saltwater intrusion has the greatest effect on coastal aquifers (Van Dijk et al. 2009).

- Change in oxygen concentrations:

  Dissolved oxygen in surface water resources plays a key role in natural stream purification processes that need optimal dissolved oxygen levels (usually > 5.0 mg l^{-1}) in order to provide for aerobic life forms. The concentration of dissolved oxygen depends highly on the water temperature. Usually, higher water temperatures cause significant decrease in dissolved oxygen concentrations (Figure 6.3).

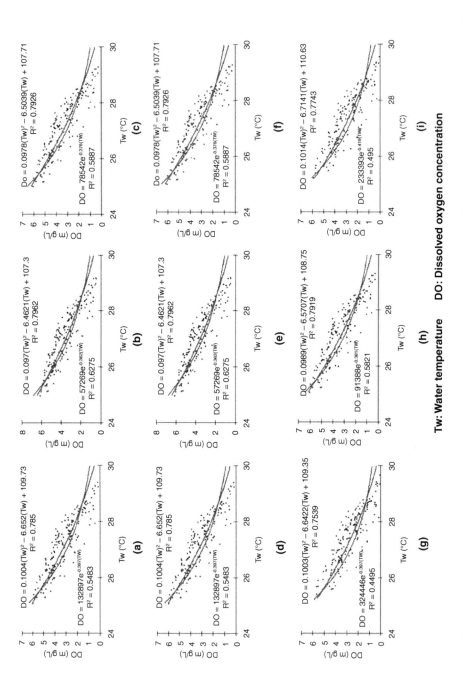

FIGURE 6.3 Correlation between water temperature and dissolved oxygen concentration under different climate change scenarios: (a) S1 (b) S2 (c) S3 (d) S4 (e) S5 (f) S6 (g) S7 (h) S8 (i) S9 (details about these scenarios in Danladi Bello et al. 2017.)

Furthermore, the aerobic metabolic rates of living organisms in the surface waters and respiration by bacteria increase with water temperature. Therefore an increase in water temperature can reduce the dissolved oxygen supply (due to solubility decrease) and can lead to significant increases in the biological oxygen demand (Rounds et al. 2013).

6.4 EFFECTS OF CLIMATE CHANGE ON GROUNDWATER

Groundwater is a key element in the hydrological cycle and is an important natural resource providing water for agricultural, industrial, and domestic purposes worldwide (WWAP 2009). Climate change has significant effects on the various components of the hydrological cycle such as evaporation, rainfall, and evapotranspiration. These effects contribute to an important alteration in the water present in rivers, lakes, oceans, etc. Groundwater recharge, storage, and quality are directly impacted by changes in the rate of precipitation and evapotranspiration (Figure 6.4).

6.4.1 Effects of Climate Change on Groundwater Recharge and Storage

Quantifying groundwater recharge and storage is challenging, especially given the absence of long-term observations of climate variables such as rainfall and evapotranspiration. The hydrologic process of recharge and storage is affected by many factors including climate variables (e.g. temperature and rainfall), soil properties (e.g. soil water content), relief properties (e.g. slope), shallow aquifer properties, vegetation cover, and agricultural water practices. These factors can lead to non-linear responses in recharge rates to changes in rainfall or evapotranspiration (Ng et al. 2010). Accordingly, the groundwater recharge rates will be the result of the interaction of all of these factors at the level of the monitored area. Good quantifying for all of these factors is essential for accurately estimating future groundwater recharge (i.e. estimation with high accuracy). However, natural climate cycles and anthropogenic factors (e.g. increase in groundwater pumping) may decrease the accuracy of estimation (Melillo et al. 2014).

Rainfall is the main contributor for groundwater recharge and storage, especially in the regions already suffering from water stress. Generally, longer-term decrease in rainfall would highly contribute to a significant decrease in groundwater recharge and storage. On the other hand, climate change is expected to change the timing, intensity, and distribution of rainfall. Evapotranspiration, another contributor for groundwater recharge and storage, is also expected to increase in the future, although the largest evapotranspiration changes are expected to occur in the summertime when recharge rates are already naturally limited. Together with expected increases in air temperature, these changes are expected to contribute to significant changes in hydroclimate: wet regions will become wetter; dry regions will become drier; and

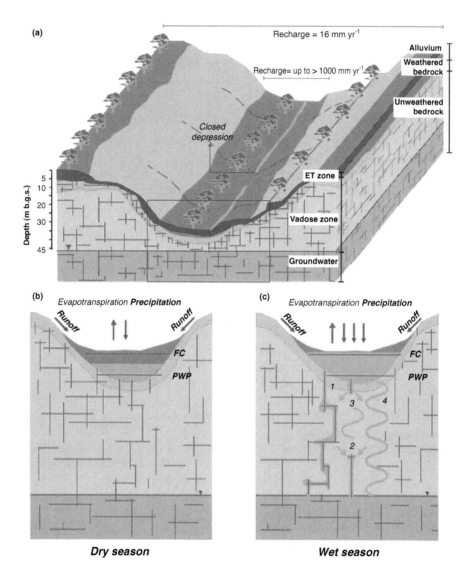

FIGURE 6.4 Impacts of precipitation and evapotranspiration on groundwater recharge (adapted from Manna et al. 2019.)

rainfall amounts will increase during the winter and decrease during the summer. Each of these factors has the potential to alter the timing, amount, and location of groundwater recharge. Without consideration of anthropogenic effects, each change to recharge the volume will be balanced by changes in groundwater storage and natural discharge. The relative ratios of change in storage and discharge depend mainly on the local hydrogeologic conditions.

The sensitivity of a given groundwater to recharge fluctuations and the lag times required for downgradient storage responses to those changes are dependent on many factors, including especially the hydraulic characteristics of the affected groundwater system, the groundwater depth, and the length of the flow path between the recharge point and the supply well (Waibel et al. 2013).

In the coming years, the expected warming climate (i.e. increases in air temperature and evapotranspiration) especially in the semi-arid and arid regions that are already showing a moisture deficit and insignificant groundwater recharge (IPCC 2013) could have a comparatively limited impact on groundwater recharge. Contrarily, in the humid regions, the expected increases in rainfall amounts can limit the role of air temperature and evapotranspiration in influencing infiltration. On balance, the additional rainfall is likely to either effectively be canceled by the concurrent increase in evapotranspiration or drive a small increase in net annual groundwater recharge (Meixner et al. 2016).

The most suitable method for quantifying climate change effects on groundwater recharge is the use of a series of linked numerical computer models (e.g. MODFLOW model). In this method, firstly long-term estimations of future climate variables such as air temperature and rainfall derived from large-scale atmospheric models are scaled-down to the level of the monitored region. Then, the downscaled values of climate variables are exploited as an input boundary condition for numerical models of surface and subsurface water flow. This method contributes highly to quantitatively identify how a groundwater recharge is likely to respond to a potential future climate condition. Moreover, this method is helpful for developing the appropriate strategies for minimizing the negative consequences of climate change on groundwater storage. As revealed by several international studies (e.g. Waibel et al. 2013; Melillo et al. 2014), avoiding further groundwater storage losses, and the hydrologic and water quality consequences associated with such losses, is a critical priority for water managers. Quantifying groundwater recharge and storage will not only provide the data required to make correct choices about water supply, but will also help researchers track closely related changes in recharge and flow discharge (hydrologic processes that are intimately connected to storage).

6.4.2 Effects of Climate Change on Groundwater Quality

Although groundwater quality is a major limiting factor for its use, it is observed, however, that few works have taken any interest in climate change effects on groundwater quality. In contrast, several studies have investigated groundwater recharge and storage (e.g. Ng et al. 2010). Groundwater quality depends on several factors. The major factors include geological factors, climate factors, and anthropogenic factors.

- Geological factors:

    These factors refer mainly to the properties of rocks and especially properties of soils (e.g. temperature, humidity, porosity, texture, pH, organic matter, infiltration, cations, microorganisms, etc.). For example, unlike non-porous soils, porous soil may better improve groundwater quality due to the better purification, which depends on soil thickness, soil texture, chemical composition of soil, and the rate of water percolation (infiltration). In addition, clay soils and rocks that contain hematite and magnetite contribute positively to the adsorption of viruses (Bitton and Harvey 1992). Furthermore, groundwater quality varies as a function of its chemical composition, affected by the solubility of the soil it passes through and by groundwater depth (Sonkamble et al. 2012). Usually, water will have low salinization levels if it passes through poorly soluble soils. Contrarily, water will show high salinization levels if it passes through soluble soils (e.g. carbonate rocks).

- Climate factors:

    Climate variables have significant effects on groundwater quality (Idoko 2010). For example, rainfall has a direct effect on groundwater recharge rate, which in turn affects the concentration of several ions elements in the water (Makwe and Chup 2013). Under normal climate conditions (e.g. there is not a heavy rainfall), the retention of microorganisms by the soil is ideal and hence, the presence of these microorganisms is insignificant. In contrast, with extreme climate events (e.g. there is heavy rainfall) the groundwater recharge is negatively affected (Tornevi et al. 2014), leading to the dissemination of pathogenic microorganisms to areas in which they were previously absent.

- Anthropogenic factors:

    These factors refer to the industrial activities (e.g. emissions of toxic chemical elements, processing residues and waste) and agricultural practices (e.g. use of fertilizers) that can contribute to the contamination of groundwaters (Huang et al. 2012). For example, an intensively irrigated agriculture leads often to an increase of nitrate concentrations, which are present in low concentrations in groundwater (Widory et al. 2004). Currently, the progressive increasing in nitrate concentrations in groundwater is a serious challenge in several parts of the word.

Groundwater quality will respond to climate change and associated anthropogenic activities due to the effects of recharge, discharge, and land use on groundwater systems (Earman and Dettinger 2011).

Groundwater quality is a major limiting factor for its use. Accordingly, the sustainability of water supplies under future climate conditions depends highly on groundwater quality. Climate change has various effects on groundwater quality (e.g. water temperature, ions concentration, salinization, etc.). Recent research (e.g. Earman and Dettinger 2011) suggests that groundwater temperatures are sensitive to climate warming. Warmer groundwater temperatures could significantly affect the concentration of ions in water. The expected change in climate has the potential to affect the physical and geochemical processes that can directly or indirectly contribute to significant effects on the groundwater quality (Dragoni and Sukhija 2008). Climate change and variability can affect the groundwater quality directly through aquifer replenishment, indirectly through a change of land use that leads to various recharge patterns and pollutants loading, shifts in biogeochemical processes and concentrated channels in the soil and unsaturated zone (Table 6.1).

Many of the dissolved ions in the groundwaters are derived from the interactions between rock formations and groundwater. Climate change has a high potential to alter the timing of these interactions and has the ability to change the chemical conditions during these interactions. Consequently, a significant degradation in groundwater quality could occur. Furthermore, the spatiotemporal variability in rainfall patterns will lead to substantial infiltration events. Large, pore-water salt reservoirs in the vadose zone, mainly chloride and nitrate, will likely be flushed into many aquifers leading to increased groundwater salinization (Sugita and Nakane 2007; Gurdak et al. 2007).

On the other hand, due to the significant impacts of recharge rates on groundwater recharge, the modeling of climate change effects on groundwater quality requires not only the accurate downscaled projections from the general circulation models, but also reliable estimations of groundwater recharge.

6.4.3 Effects of Sea-Level Rise on Groundwater

The rise in sea-level is one of the most significant processes that has resulted from the climate change effects. This mechanism has the potential to change the horizontal and vertical positions of the water table and the slope and direction of the hydraulic pressure gradient, which would, in turn, modify the position of the freshwater–saltwater interface. These changes induce the sea intrusion process that can decrease the volume of freshwater and increase

TABLE 6.1
Pathways That Climate Change Affects Components of Groundwater System

Climate Scenario	Pathway	Potential Effects
Changes in temperature and rainfall	Recharge	Modification in leachate transport
Warmer temperature	Land use	Favorable condition for arable crops
Warmer temperature	Biochemical processes	Enhanced denitrification
Heavy rainfall	Concentrated paths	Microbial contamination
Warmer temperature and less rainfall	Concentrated paths	Crack formation in the soils

the salinization of coastal aquifers. Such changes can result in an important water quality effect in coastal-area aquifers used for supply purposes. The direct effects of sea-level rise (saltwater intrusion and saltwater inundation) on groundwater are likely to be largest in settings with very low topographic relief and very low hydraulic gradients between freshwater and marine water (Ferguson and Gleeson 2012). Although climate-driven changes in sea-level position will increase the potential for seawater intrusion into coastal aquifers, poorly managed near-shore groundwater pumping is likely to continue to be the dominant factor driving intrusion in several coastal regions. Increases in near-shore pumping rates in response to climate change could further affect coastline areas that are already suffering from a saltwater intrusion problem.

6.5 CONCLUSIONS

Climate change causes significant effects on water resources availability and management. It also causes significant effects on groundwater recharge and storage. Since groundwater is a key element in the global hydrologic cycle, more attention should be paid to the effects of climate change on recharge. Furthermore, climate change affects the physico-chemical quality of water resources (surface waters and groundwaters). Climate change affects the quality of water resources through various ways such as bio-chemical processes, acidification, salinization, etc. Generally, it is expected that climate change will reduce the physico-chemical quality of water resources. Development of a tool to assess the global change effects on water resources (based mainly on climate models) would greatly help decision makers and water resources companies in their long-term planning and in the development of adaptation strategies. This is highly recommended for the regions already suffering from water stress.

7 Potential for Soils to Mitigate Climate Change

7.1 INTRODUCTION

Soils contain the largest store of organic carbon with about 2,400 Pg (Baldock et al. 2012). This large carbon store is always exposed to decomposition and is impacted by several biological processes that produce or consume many greenhouse gases (Figure 7.1), especially carbon dioxide (CO_2), nitrous oxide (N_2O), and methane (CH_4). Accordingly, depending on their management, soils have the capacity to either increase or decrease the emissions of greenhouse gases in the atmosphere.

Suitable management of soils can play a significant role in climate change mitigation by storing carbon (i.e. carbon sequestration) and reducing atmospheric greenhouse gas emissions. In contrast, unsuitable management of soils can contribute negatively to climate change by carbon emission in the atmosphere (in the form of carbon dioxide). Under normal carbon cycle (i.e. all components of the carbon cycle remained constant), a 1% change in the amount of carbon contained in soils would equate to about 8 ppm change in the amount of carbon dioxide contained in the atmosphere (Sanderman et al. 2010). The steady conversion of grassland and forestland to cropland and grazing lands has resulted in significant losses of soil carbon in many regions of the world. However, by restoring degraded soils and adopting soil conservation options, there is high potential to reduce the emission of greenhouse gases from agriculture, enhance carbon sequestration, and mitigate climate change. In this chapter, we will focus on how soil can mitigate climate change (i.e. carbon sequestration) and then we will discuss the major soil management practices that can enhance carbon storage in the soils.

7.2 SOIL CARBON SEQUESTRATION TO MITIGATE CLIMATE CHANGE

Soil carbon sequestration is a long-term process in which atmospheric carbon dioxide (CO_2) and other forms of carbon are stored in the soil. Accordingly, the process is considered as an elusive climate mitigation tool (Lal 2001). Figure 7.2 shows an overview of carbon capture and storage.

Several mechanisms contributed to the carbon dioxide storage. These mechanisms include biological, geological, mineral, and ocean mechanisms. Table 7.1 shows the worldwide CO_2 storage potential of these mechanisms. The biological

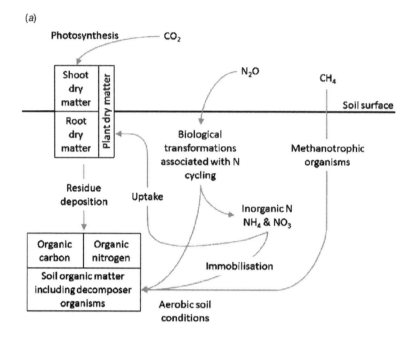

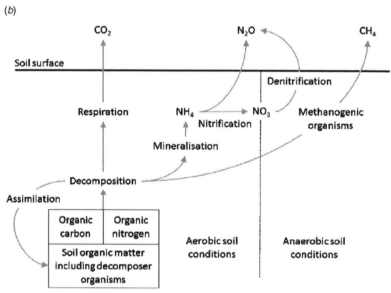

FIGURE 7.1 Soil biological processes that affect consumption (a) or emissions (b) of greenhouse gases (adapted from Baldock et al., 2012).

Potential for Soils to Mitigate Climate Change

Carbon Capture and Storage (CCS)

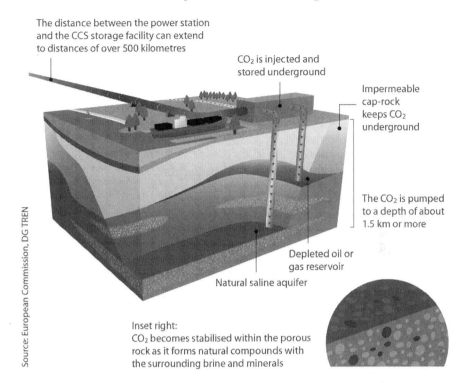

FIGURE 7.2 An overview of carbon capture and storage.

TABLE 7.1
Worldwide Storage Potential for CO_2
(adapted from Herzog et al. 1997)

Sequestration Process	Storage Capacity (Billion Tons of CO_2)
Biological sequestration	Unknown
Geological sequestration	320–10,000
Mineral sequestration	> 30,000
Ocean sequestration	>100,000

mechanisms (called also terrestrial mechanisms) refer to the intervention of plants and micro-organisms in the storage of carbon dioxide in vegetative biomass and in soils. The geological mechanisms refer to storage of carbon dioxide in various geological formations. In the mineral mechanisms, carbon, in the form of carbon dioxide can be stored in stable carbonate mineral forms (Saran et al. 2017). Finally, the ocean mechanisms refer to the uptake of carbon dioxide by the oceans. In the following sections, we discuss in detail these various mechanisms.

7.2.1 BIOLOGICAL SEQUESTRATION

In biological sequestration, the carbon element is driven by photosynthesis in which plants take up carbon dioxide and water and turn it into reduced carbon products such as cellulose. Yearly, the photosynthesis process has the potential to fix more than 90 Gt of carbon. Most of this potential is returned through respiration and decomposition mechanisms; however, recently mid latitude forests have been net sinks for carbon dioxide (Izaurralde and Rice 2006). Several plants, especially the trees, sequester carbon dioxide during growth phase. Therefore, forestation can contribute positively to carbon sequestration (either in vegetative biomass or in soils). Mid latitude forests have the potential to fix more than 60 tons of carbon per hectare above ground and about 100 tons of carbon per hectare below ground. Hence, planting forests can be an effective tool to mitigate climate change (Lal 1999).

On the other hand, ocean fertilization is also a biological sequestration option. In this option, fertilization aims to remove carbon dioxide directly from the atmosphere by stimulating the natural biological processes in the surface oceans (Herzog et al. 2000). This option spurs biomass growth in areas that are low in productivity due to lack of critical nutrients. However, it is limited in capacity, and has raised various environmental risks such as changes in the natural food chain. Concerns regarding the efficacy and environmental risks of fertilization should be resolved first by a more thorough synthesis of disciplinary knowledge in the aquatic sciences, including input from ecologists and limnologists. Then, if the arguments are still not persuasive enough for sound decision making, limited scientific testing of carbon sequestration options may be justified (Lancelot et al. 2000).

7.2.2 GEOLOGICAL SEQUESTRATION

This sequestration involves the separation and capture of carbon dioxide at the emissions source followed by storage in various geological formations (IPCC 2013). The major geological formations include active oil fields, coal beds, depleted oil and gas reservoirs, deep saline aquifers, and mined salt domes and rock caverns (Figure 7.3).

Geological carbon sequestration can occur by physical and by chemical mechanisms. The physical mechanisms involve trapping carbon dioxide within a cavity in the rock underground. Two groups of cavities can be distinguished: large man-made cavities (e.g. caverns and mines) and structural traps in depleted oil and gas reservoirs and in aquifers. On the other hand, the chemical mechanisms of trapping carbon dioxide involve transforming the carbon dioxide to another substance in the ground. This transformation is assured by several

Potential for Soils to Mitigate Climate Change

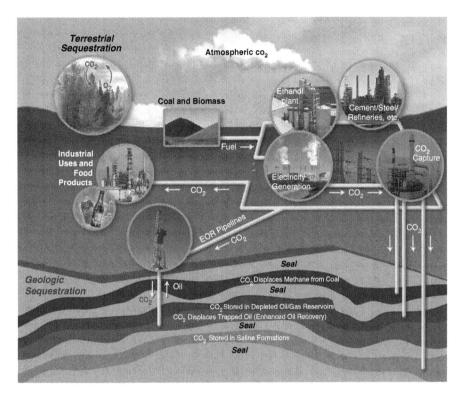

FIGURE 7.3 An overview of geological sequestration.

chemical methods, including dissolving carbon dioxide in underground water or reservoir oil; decomposing carbon dioxide into its ionic components; and locking carbon dioxide into a stable mineral precipitate (Lackner et al. 1997).

For example, in Texas (USA), 20 million tons of carbon dioxide are stored in tertiary oil recovery. In this case, the main advantage of the geological sequestration option consists that most of the injected carbon dioxide will be immobilized by either physical or chemical absorption on the coal surface. On the other hand, in other reservoirs such as aquifers, the injected carbon dioxide will likely exist as a supercritical phase without being fully dissolved for thousands of years and that creates a potential problem of a higher leakage rate (Johnson et al. 2001).

Deep aquifers are the best geological formations for carbon storage. For example, in the USA, in 2010, the storage capacity of 15 deep aquifers reached 500 billion tons of carbon dioxide (Brennan et al. 2010). Several international studies reported that injection into deep aquifers is technically and economically feasible. Worldwide, there are a large number of sites where carbon dioxide can be stored in deep aquifers. For the application of the geologic sequestration, both near surface and in depth monitoring systems should be developed to minimize carbon dioxide leakage from underground.

7.2.3 MINERAL SEQUESTRATION

The increasing carbon dioxide concentration in the Earth's atmosphere, especially caused by fossil fuel combustion, makes it difficult to develop a physical barrier that completely prevents carbon dioxide from returning to the atmosphere (Goldberg et al. 2001). Mineral sequestration (Figure 7.4) involves the dissolution of minerals and subsequently carbonation of dissolved minerals.

The carbonation process could be assured directly or indirectly (Saran et al. 2017). A general overview of direct and indirect mineral carbonation (adapted from Bobicki et al. 2012) is presented in Figure 7.4.

- Direct carbonation:

  Direct carbonation of calcium (Ca) and magnesium (Mg)-based minerals in a single process step can be assured through two main routes: direct dry gas-solid reaction and direct aqueous carbonation:

  ○ Direct dry gas-solid reaction:

  The reaction between carbon dioxide (CO_2) and calcium/magnesium-based silicate minerals (such as olivine (($Mg, Fe)_2 SiO_4$) and serpentine ($Mg_3Si_2O_5(OH)_4$)) which forms stable magnesium carbonate ($MgCO_3$) and calcium carbonate ($CaCO_3$) is a good example for this direct dry gas-solid reaction. The general reaction of calcium/magnesium-based silicate minerals with CO_2 is as follows (Maroto-Valer et al. 2005):

  $$(Ca/Mg)_x Si_y O_{x+2y} + xCO_2 \rightarrow x(Ca/Mg)CO_3 + ySiO_2 \quad (7.1)$$

  For magnesium-based minerals, the exothermic reaction of these minerals with CO_2 is as follows:

  $$Mg_3Si_2O_5(OH)_4(s) + 3CO_2(g) \rightarrow 3MgCO_3(s) + 2SiO_2(s) + 2H_2O(l) \quad (7.2)$$

  where s: solid; g: gaz; and l: liquid. Advantages and disadvantages of the dry gas-solid reaction are summarized in Table 7.2.
  Getting high carbonation reaction rates requires a significant increase in temperature and carbon dioxide pressure (Zevenhoven and Kohlmann 2002). However, the potential increase of the temperature is thermodynamically restricted because the chemical equilibrium favors gaseous carbon dioxide over solid-bound carbon dioxide at high temperatures due to entropy effects. Consequently, the maximum temperature at which the direct carbonation occurs spontaneously depends highly on the carbon dioxide pressure and the type of mineral (Table 7.3). Activation of the feedstock by heat treatment can highly ameliorate the carbonation rate; however, it is quite energy-consuming (Zevenhoven and Kohlmann 2002; Zevenhoven and Teir 2004). On the other hand, it is important to mention that during the surface carbonation process, a thin layer of the carbonates forms as a diffusion barrier to both the

Potential for Soils to Mitigate Climate Change 73

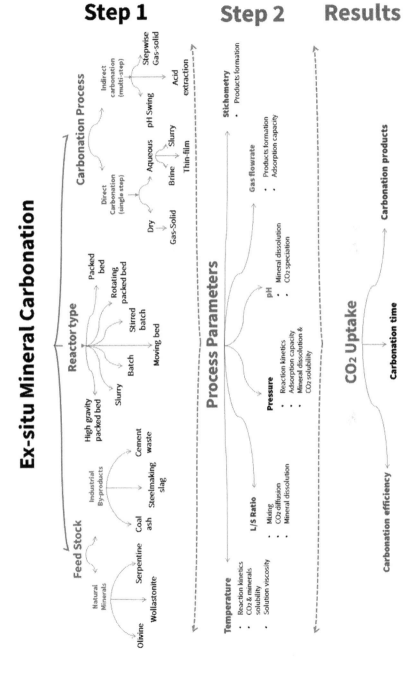

FIGURE 7.4 An overview of mineral sequestration.

TABLE 7.2
Advantages and Disadvantages of the Dry Gas-Solid Reaction

Advantages	Disadvantages
Simple process design	Very slow process
A better ability to apply the reaction	Depends on temperature
Heat generated by the carbonation reaction	Depends on carbon dioxide pressure

TABLE 7.3
Maximum Allowable Reaction Temperatures at Corresponding Pressure for a Number of Minerals (Adapted from Lackner et al. 1995)

Mineral	Maximum Temperature (°C)	CO_2 Pressure (Bar)
Calcium oxide (CaO)	888	1
Magnesium oxide (MgO)	407	200
Calcium hydroxide ($Ca(OH)_2$)	888	1
Magnesium hydroxide ($Mg(OH)_2$)	407	200
Wollastonite ($CaSiO_3$)	281	1
Forsterite (olivine) (Mg_2SiO_4)	242	1

outward diffusion of water (H_2O) and the inward diffusion of CO_2, decreasing the carbonation rate.

- **Direct aqueous carbonation reaction**

    In this reaction, the calcium/magnesium-based silicate minerals are carbonated in an aqueous suspension (O'Connor et al. 2000). Here, the contribution of water (H_2O) greatly contributes to the enhancement of the carbonation rate. Carbonation potentials of some natural minerals through direct aqueous mineral reaction are summarized in Table 7.4.

    In this reaction, the carbonation is assured through three successive steps:

*Step 1: the dissociation of CO_2 in water

In this step, CO_2 dissociates to bicarbonate and H^+, resulting in a mildly acidic environment with HCO_3^- as the dominant carbonate species (Equation 7.3):

$$CO_2(g) + H_2O(l) \rightarrow H_2CO_3(aq) \rightarrow H^+(aq) + HCO_3^-(aq) \quad (7.3)$$

*Step 2: the Ca/Mg leaching

In this step, Ca/Mg leaches from the mineral matrix accelerated by the available protons (Equation 7.4):

TABLE 7.4
Carbonation Potential of Natural Minerals Through Direct Aqueous Mineral Carbonation (Adapted from O'Connor et al. 2005)

Rock	Mineral	Formula	Carbonation Potential
Serpentine	Antigorite	$Mg_3Si_2O_5(OH)_4$	2.1
Serpentine	Lizardite	$Mg_3Si_2O_5(OH)_4$	2.5
Olivine	Forsterite	Mg_2SiO_4	1.8
Pyroxene	Augite	$CaMgSi_2O_6 + (Fe, Al)$	2.7
Ultramafic	Wollastonite	$CaSiO_3$	2.8

$$Ca/Mg - silicate(s) + 2H^+(aq) \rightarrow (Ca/Mg)^{2+}(aq) + SiO_2(s) + H_2O(l) \quad (7.4)$$

*Step 3: Ca/Mg precipitation

In this final step, Ca or Mg carbonates precipitate according to the following equation (Equation 7.5):

$$(Ca/Mg)^{2+}(aq) + HCO_3^-(aq) \rightarrow (Ca/Mg)CO_3(s) + H^+(aq) \quad (7.5)$$

In the aqueous mineral carbonation process, the carbonation rate increase upon a temperature increase is counteracted by a solubility decrease of CO_2 in the water (Gerdemann et al. 2002). A significant increase of the carbonation rate could be accomplished by, for example, an increase of the specific surface area, removal of the SiO_2-layer and decreasing the $(Ca, Mg)^{2+}$-activity in solution (Chizmeshya et al. 2004).

- Indirect carbonation:

In this carbonation, the dissolution of reactive mineral ions from the feedstock and the carbonation of dissolved mineral ions take place separately in two different reactors (O'Connor et al. 2005). The indirect mineral carbonation involves two successive major stages. The first stage refers to the extraction of reactive elements (Mg^{2+}, Ca^{2+}) from calcium/magnesium-based minerals using usually acids. Several strong acids (e.g. HCl, H_2SO_4, and HNO_3) are used for the dissolution of silicate rocks (Wang and Maroto-Valer 2011). The second stage refers to the reaction of the extracted elements with CO_2 in either the gaseous or aqueous phase. Pure carbonates can be produced using indirect ways, because of the removal of impurities in previous carbonate precipitation stages. On the other hand, as high carbonation efficiency was only assured after a long time, indirect mineral carbonation is not viable on an industrial scale. Over recent years, in order to enhance the efficiency of mineral dissolution and recovering and re-using additives, Maroto-Valer et al. proposed a pH-swing CO_2 mineralization process using ammonium salts (Wang and Maroto-Valer 2011). At 100 °C, 1.4 M aqueous solution NH_4HSO_4 was found to extract 100% Mg from serpentine in 3 hours.

7.2.4 OCEAN SEQUESTRATION

The oceans are considered as the biggest store for carbon dioxide emissions. Currently, the world's oceans have the potential to uptake a great amount of carbon per year (Figure 7.5). This important uptake is not a consequence of deliberate sequestration; it occurs due to the natural chemical reactions between seawater and carbon dioxide in the atmosphere. However, these natural chemical reactions usually lead to ocean acidification.

Several marine organisms and ecosystems depend on the formation of carbonate skeletons and sediments that are vulnerable to dissolution in acidic waters. Laboratory and field measurements show that carbon dioxide-induced acidification may eventually cause the rate of dissolution of carbonate to exceed its rate of formation in these ecosystems. The effects of ocean

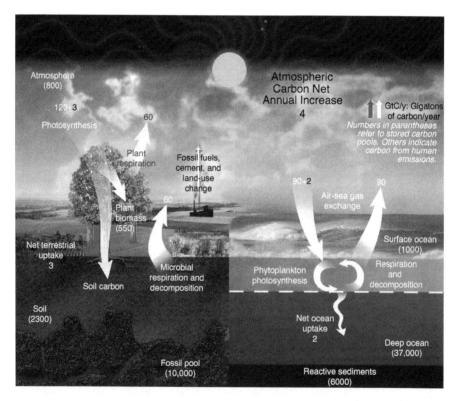

FIGURE 7.5 Diagram showing a simplified representation of the Earth's annual carbon cycle (adapted from Courtesy of Oak Ridge National Laboratory, U.S. Dept. of Energy).

acidification and deliberate ocean fertilization on coastal and marine food webs and other resources are poorly understood. Several researchers are focusing on the impacts of oceanic carbon sequestration on these significant environments. On the other hand, it is important to mention that the identified oxygen reduction in the world's oceans interior as a result of global climate change could highly affect prokaryotic chemoautotrophic processes and hence the overall carbon storage capacity of the oceans. This issue clearly matters in terms of climate change mitigation and adaptation, however is rarely if ever considered.

7.3 RECENT SOIL MANAGEMENT PRACTICES FOR INCREASING CARBON STORAGE IN SOIL

The potential of soil organic carbon sequestration is highly important in the world's degraded soils and ecosystems, estimated at 1216 Mha, and agricultural soils, estimated at 4961 Mha (Lal 2002). These soils have lost an important part of their original organic carbon store, and have the ability to sequester carbon by converting to a restorative land use and adopting recommended management practices (Table 7.5). In this section, we discuss mainly some recent soil management practices that can help to increase levels of carbon in soils, and hence can help to mitigate climate change.

- Agroforestry:

  Agroforestry system refers to the cultivation of trees and crops in interacting combinations. This intercropping has various benefits and services such as increasing crop yields, decreasing food insecurity, enhancing environmental services, and resilience of agroecosystems (Ajayi et al. 2011). Furthermore, as revealed by several studies (e.g. Nair et al. 2009; Schoeneberger et al. 2012; Cubbage et al. 2013), this system is recognized as an integrated strategy for sustainable soil use aside from its contribution

TABLE 7.5
Traditional and Recommended Soil Management

Traditional Soil Management	Recommended Soil Management
Biomass burning and residue removal	Residue returned as surface mulch
Conventional tillage and clean cultivation	Conservation tillage, no till and mulch farming
Bare/idle fallow	Growing cover crops during the off-season
Monoculture	Crop rotations with significant diversity
Soil fertility mining	Judicious application of off-farm input
Intensive application of chemical fertilizers	Nutrient management with compost
Intensive cropping	Integrating tress and livestock with crop production
Surface irrigation method	Furrow or sub-irrigation method
Indiscriminate application of pesticides	Pest management
Cultivating marginal soils	Restoration of degraded soils through land use change

to climate change mitigation. Globally, an estimated 700, 100, 300, 450, and 50 Mha of soil are used for tree intercropping, multistrata systems, protective systems, silvopasture, and tree woodlots, respectively (Nair 2012). Compared to monoculture practice, the agroforestry system has a great potential to increase soil organic carbon sequestration. Indeed, the root system of incorporated trees, in particular, enhances soil characteristics and can result in greater net carbon sequestration (Young 1997). Due to their important extension, the root systems of incorporated trees in agroforestry can interact well with soil particles, especially in the deep soil layers. It can exploit zones of rich, localized, supplies of water and nutrients. Accordingly, root-derived carbon is more likely to be stabilized in the soil through various physicochemical interactions with soil particles than shoot-derived carbon (Rasse et al. 2005). For example, the relative root contribution of European beech to soil organic carbon was 1.6 times that of shoots (Scheu and Schauermann 1994). Similarly, in croplands, total root derived carbon contributed from 1.5 to more than 3 times more carbon to soil organic carbon than shoot-derived carbon (Johnson et al. 2006). On the other hand, the beneficial ability of tree roots to uptake nutrients from deep soil layers (e.g. below the tree rooting zone) leads to significant enhancement of plant growth by subsequent increase in nitrogen (N) nutrition. This increase leads in turn to significant increase in soil organic carbon sequestration. Similar, mixed plantings with nitrogen fixing trees leads to higher biomass production and, therefore, soil organic carbon sequestration and pools especially in deep soil layers as nitrogen promotes humification rather than decay.

- Cultivation of perennial crops:

  Cultivation of perennial crops has several benefits. The major benefits include soil organic matter increasing, hence higher soil fertility, soil tillage reduction, hence, higher resistance to soil erosion, and soil properties improving (e.g. soil structure amelioration, soil water content increase, and soil salinization decrease). All these benefits have a significant positive impact on crop growth. Overall, these benefits can help to mitigate the negative effects of climate change on soils and on crops (Harvey et al. 2014). Also, perennial crops require less fossil fuel inputs than annual crops. Several studies (e.g. Cox et al. 2013; de Leeuw et al. 2014) have revealed that perennial crops have reduced the need for fuel, fertilizer, and pesticides. Perennial crops have higher rates of carbon sequestration and offer various benefits to the farm, the farmer, and the surrounding ecosystem. Due to their high age (compared to annual crops), several perennial crops invest in deep, extensive root systems which are efficient in scavenging for nutrients, water, and putting carbon in to soil. The amount of carbon retained by soils is affected highly by management practices, with those that lead to decrease soil disturbance and for enhanced crop persistency having the greatest benefits on carbon sequestration. Annual staple crops occupy the great majority of cropland: roughly 1 billion ha of the world's 1.5 billion ha total cultivated land (Pimentel et al. 2012). Perennial crops have an important potential to perennialize some of this annual cropland. Carbon sequestration rates for perennial staple crops vary from 1 to 7 Mg ha^{-1} yr^{-1}(Harvey et al. 2014). These rates are highly variable, and data are sorely lacking. Nonetheless the carbon sequestration potential of perennial staple crops is clear.

- Conservation tillage:

  Conservation tillage and adding organic materials to soil have significant positive impact on soil carbon increasing (Powlson et al. 2012). Contrary to traditional tillage system, the specific objective of conservation tillage systems consists of balanced conservation of soil, water, and energy resources. This conservation is assured through the decrease of tillage intensity and retention of crop residue. Technically, the conservation tillage system refers to the planting, growing, and harvesting of crops with very low disturbance to the soil surface. According to a study performed by Dinnes (2004), with the conservation tillage system, at least 30% of the soil surface could be covered with crop residue/organic residue following planting. The system stimulates microbial decomposition of soil organic matter, which leads to the emissions of carbon dioxide to the atmosphere. Consequently, decreasing the amount of tillage enhances sequestration of carbon in the soil. Advancements in weed control techniques and farm machinery allow several crops to be grown with minimum amount of tillage (Smith et al. 2008). The conservation tillage system has several benefits. The major benefits include the following: enhancing the capacity of soil to store or sequester carbon; ameliorating soil water infiltration (this leads to reducing soil erosion and nitrate runoff); decreasing nutrients leaching (as a result of greater amounts of soil organic matter); and reducing evaporation and increasing soil water content, which can enhance crop yields in drought years (Suddick et al. 2010). However, some minor disadvantages were identified (e.g. adoption of conservation tillage in humid, cool soils would primarily affect the distribution of soil organic carbon in the profile, unless carbon inputs were increased). Worldwide, several tillage networks were established to promote the use of a conservation tillage system instead of traditional tillage. For example, the Africa Conservation Tillage Network was established in 1998 to make more effective use of natural and human resources, promote conservation tillage practice, and reduce degradation of agricultural areas (Abrol et al. 2005).

- Enhanced mineral weathering in soils to sequester carbon dioxide:

  This practice refers to the acceleration of silicate rocks weathering on land in order to get carbonate rocks, consequently, fixing carbon dioxide from the atmosphere (Köhler et al. 2013). This practice has recently been developed as a method of mitigating anthropogenic climate change. The natural transformation of silicate rocks to carbonate rocks contributes highly to the consumption of atmospheric carbon dioxide and is a key component of geochemical cycles and of the climate system on long timescales (Berner and Kothavala 2001). This natural transformation can be artificially accelerated by applying finely-divided silicate rocks such as olivine to soils, with the carbonate so fixed being stored in the soil as mineral carbonate and, then, exported in drainage waters to rivers and the oceans. The acceleration of silicate rocks weathering has several benefits. For example, it can help shape the Earth's surface, regulate global and chemical cycles, and even determine nutrient supply to ecosystems. However, the effects of this acceleration on ecological systems such as land and ocean need further evaluating to avoid any negative consequences (Hartmann et al. 2013). Weathering of silicate minerals depends strongly on the mineral dissolution rates that occur (Moosdorf et al. 2014). These rates depend highly on air temperature, mineral saturation, pH and mineral surface area (Pokrovsky and Schott 2000). Due to these various factors, usually, the mineral dissolution rates are determined from laboratory flow-through dissolution experiments which seek to bridge this observational discrepancy by using columns of soil returned to the laboratory from a field site. Furthermore, the dissolution rates provide critical information for the assessment of enhanced weathering including the expected surface-area and energy requirements (Renforth 2012).

- Enhanced storage of carbon in subsoil horizons:

  This practice refers mainly to the stabilization of organic carbon in the subsoil horizons. This stabilization has received much focus recently due to its relevance in controlling the global carbon cycle. The subsoil horizons have an important potential to soil organic carbon sequestration. Carbon sequestration in these horizons contributes to more than half of the total carbon stocks in soil. Several international studies (e.g. Six et al. 2002; Watts et al. 2005) have reported that, in the subsoil horizons the major factors controlling soil organic carbon stabilization include soil properties (especially clay minerals), environmental factors, soil management factors, and temperature and soil water content dynamics. Enhancing the storage of organic carbon in subsoil horizons needs many requirements. The main requirements include (i) a mechanism for organic carbon emplacement at depth, and (ii) confidence that this carbon will stay in the soil (i.e. be stabilized), and will not be degraded by soil biota. The suitable option of direct organic carbon emplacement is to enhance the natural process of migration of carbon from topsoil to subsoil; this is dominated by leaching of dissolved organic carbon. Leaching of dissolved organic carbon is known to be greater under forests than grassland and arable land. Minerals deeper in the soil profile, such as iron oxides, have a large capacity to stabilize this carbon (Mikutta et al. 2006). On the other hand, the lower dynamics of temperature and soil moisture in subsoil horizons may thus enhance or reduce organic matter mineralization, while low nutrient availability is a more common limiting factor. Spatial distribution of organic matter may determine the likelihood of its stabilization at long time scales, which may be most related to absence of energy-rich material needed for decomposition. Therefore a suitable option for increasing carbon stocks in subsoil horizons may be the addition of highly stable organic matter such as biochar or highly aliphatic material.

7.4 CONCLUSIONS

In the present chapter, the importance of soils as a source and sink for atmospheric carbon is well established. In addition, in this chapter, various recent soil management practices such as agroforestry, cultivation of perennial crops, and reducing tillage, that have a great potential to sequester carbon and reverse the carbon enrichment of the atmosphere, were discussed with details. However, the potential of soil carbon sequestration may be restricted in practice due to landscape variability and the effects of atmospheric change and climate warming. Regardless of these restrictions, the benefits of increasing soil organic carbon sequestration even on a limited basis are so important for maintaining and enhancing agricultural productivity, as well as supporting vital ecosystem services, and should be supported through advanced research initiatives.

8 Climate System Modeling

8.1 INTRODUCTION

The responses of the climate system to anthropogenic activities are so complex due to several factors such as the non-linearity of many processes that define the climate system and the different response times of the various components to a given perturbation (Goosse et al. 2010). As a consequence, the only tool available to evaluate these responses is by using numerical climate models. These models are a perfect tool to understand perfectly different climate processes, to create plausible-scenarios, to understand past climatic changes, to reflect present climatic changes, and to propose suitable strategies for the future (Washington and Parkinson 2005). Generally, a climate model is a mathematical representation of the main drivers of climate change (Figure 8.1).

The overall aim of climate models is to understand various processes that can contribute to climate change. The equations used to describe these various processes are so complex. Accordingly, they require numerical resolution. Consequently, climate models provide solutions for these complex equations (Sellers 1969). The solutions start from some "initialized" state and evaluate the impacts of changes in different parts of the climate system (Trenberth 1992). For example, the climate models are subdivided into two major groups. The first group refers to the models that just cover one particular region of the world or part of the climate system. The second group represents the models that can simulate the atmosphere, oceans, and land for the whole planet. The output from these models drives forward climate science, helping researchers understand how anthropogenic activities are influencing the Earth's climate. These advances have underpinned climate policy decisions on national and international scales for the past years (Washington and Parkinson 2005). In the present chapter, we will focus on how climate models work (i.e. components of climate models, numerical resolution of basic equations, input and output data, and validation of model outputs) and then we will summarize the major strengths and limitations of current climate modeling.

8.2 COMPONENTS OF CLIMATE MODELS

A climate model is mainly used to simulate each of the different components of the climate system. Therefore, the components of a global climate model include an atmospheric model, an ocean model, a land model, and a sea ice model (Trenberth 1992).

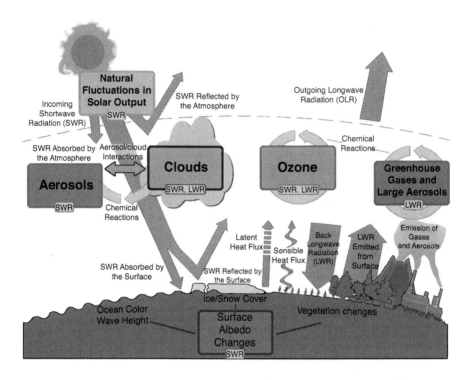

FIGURE 8.1 Main drivers of climate change (adapted from Cubasch et al. 2013.)

8.2.1 THE ATMOSPHERIC GENERAL CIRCULATION MODEL (AGCM)

The atmospheric general circulation model (AGCM) is developed and used for climate modeling. It is a mathematical model constructed around the full set of basic dynamical equations that govern atmospheric motions. These equations are developed to describe the behavior of the following parameters: velocity ($\vec{V}$), temperature (T), pressure (P), specific humidity (q), and density (ρ). These equations are written as follows (Equation 8.1 to Equation 8.7):

The velocity ($\vec{V}$) is determined based on Newton's second law as follows:

$$\vec{V} = (a)t + \vec{V}_0 = (F) + V_0 \tag{8.1}$$

where $\vec{V}$: is the velocity (m s^{-1}), a: is the object acceleration (m s^{-2}), t: is the time (s); F: is the net force applied on the object (N), m: is the mass of the object (kg), and V0 is the initial velocity at time = 0 (m s^{-1}).

The "conservation of momentum" in the atmosphere is determined based on the mass conservation equation (or continuity equation):

$$\frac{dM}{dt} = \frac{d}{dt}\int_V \rho dV = \int_V \frac{\partial \rho}{\partial t} dV \qquad (8.2)$$

where ρ: is the density field of the fluid (–); dV: is the change in the volume (m^3); dt: is the change in the time (s); and dM: is the change in the mass (kg);

The conservation of water mass in the atmosphere is also determined based on the continuity equation as follows:

$$\partial \rho q \partial t = -\nabla \rho \vec{V} q + \rho E - C \qquad (8.3)$$

where ρ: is the density field of the fluid (–); q: is the specific humidity (g kg^{-1}); t: is the time (s); $\vec{V}$: is the velocity (m s^{-1}); and E and C are evaporation and condensation respectively.

The conservation of energy is determined based on the first law of thermodynamics as follows:

$$Q = C_p dT dt - 1\rho dp dt \qquad (8.4)$$

where Q is heating rate per unit mass (kJ. Kg^{-1}); C_p the specific heat (kJ.kg^{-1}.K^{-1}); ρ: is the density of the fluid (–); dT: is the change in the temperature (°C); dt: is the change in the time (s); and dp: is the change in the pressure (Pa).

In thermodynamics, an equation of state is developed to define the state of matter under specific physical conditions (e.g. pressure, volume, and temperature). This equation is as follows:

$$PV = nRT \qquad (8.5)$$

where P: is the pressure (Pa); V: is the volume (m^3); n: is the number of moles of the gas; R: is the universal gas constant (= 8.31J K–1); and T is the temperature (°C).

Several standard approximations have to be done before the use of the equations (equation 8.1 to 8.5) in the global climate models. Currently, assuming hydrostatic equilibrium, which is a perfect approximation at the scale of general circulation models, provides a good simplification of the equations. In addition, the quasi-Boussinesq approximation states that the time variation of the density is insignificant compared to the other parameters of the mass conservation equation, filtering the sound waves. However, further equations for the liquid water content of atmospheric parcels or other variables related to clouds are usually added to the mentioned equations (equation 8.1 to 8.5). Furthermore, the effects of clouds on heating rate process leads to significant uncertainty. These effects that could not be resolved by the model grid must therefore be parameterized, introducing new terms into the equations 8.1, 8.3, and 8.4. The boundary conditions of these equations describing the interactions between the atmosphere and the other components of the climate system also need to be defined.

8.2.2 THE OCEAN MODEL

The ocean model is a mathematical model constructed and used mainly to understand and estimate the various aspects of the ocean which is considered as a complex geophysical system. Over recent years, improved understanding of the ocean and ocean models coupled with increased computer power have led to perfect representations of ocean dynamics (such as advection, diffusion, and entrainment, see Figure 8.2).

Overall, the major differences between the atmospheric model and the ocean model are summarized in Table 8.1. The main equations that govern these dynamics are based on the same principles as the equations for the atmospheric general circulation model. However, the differences include the following two points. First, unlike the atmospheric model, the equation for the specific humidity is not needed for the ocean model; however, a new equation for the salinity is needed. Second, contrary to the atmospheric model, there is no simple law for the ocean dynamic and the equation of state is expressed as a function of the pressure, the temperature, and the salinity as a long polynomial series.

Due to the solar radiation absorption process, the oceans may be warming faster. Also, the salinization of waters may be increasing faster. This process is taken into account in the ocean model through an exponential decay of the solar irradiance. The absorption of solar radiation and the salinization of oceans are mathematically expressed in the ocean model as follows:

$$dTdt = F_{sol} + F_{diff} \tag{8.6}$$

$$dSdt = F_{diff} \tag{8.7}$$

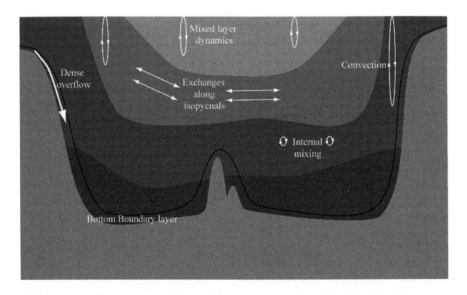

FIGURE 8.2 A simplified representation of the different processes that have to be parameterized in the ocean model (adapted from Goosse et al., 2010.)

TABLE 8.1
Major Differences Between the Atmospheric Model and the Ocean Model

Ocean Model	Atmospheric Model
Confined to only certain areas of the earth's surface	Confined to all areas of the earth's surface
Spectral representation is not used	Spectral representation is used
Smaller spatial scale of oceanic eddies compared to atmospheric eddies	Important spatial scale of atmospheric eddies
Grid resolution needs to be much finer than in atmospheric model	No problem for grid resolution
Problems for initialization, verification, and parameterization	No problem for initialization, verification, and parameterization
Higher resolution required near equator and near the poles where currents are narrower	Higher resolution not required near equator or near the poles

where dT: is the change in the temperature (°C); dt: is the change in the time (s); F_{sol}: is the absorption of solar radiation in the ocean (W m^{-2}); F_{diff}: is the diffusion term (m^2.s^{-1}); and dS: is the change in the salinization (g kg^{-1}).

In equations 8.6 and 8.7, unlike the atmospheric model, we have distinctly introduced a term to the right-hand side showing the effects of the processes at scales that cannot be added in the model. As small-scale processes have the potential to mix properties, they are usually simulated as a diffusion term (F_{diff}) through the Laplacian formulation. Due to the big different scales of ocean model grids on the vertical direction (some hundred meters) and on the horizontal direction (hundreds of kilometers), the small-scale processes in these two directions have distinct properties. Accordingly, the diffusion coefficient will differ highly in both directions. Instead of separating horizontal and vertical directions, it is more useful to use a referential that is aligned with the density surfaces. Accordingly, isopycnal (i.e. along surfaces of equal density) and diapycnal (i.e. normal to surfaces of constant density) diffusion coefficients are computed using sophisticated modules. These modules take into consideration several effects such as the stirring of the winds, the effect of density gradients, and the breaking of surface and internal waves.

8.2.3 The Land Model

The land model is a mathematical model constructed and used mainly to understand and simulate the various processes that result from the interactions between the components of Earth's land surface (such as soil, vegetation, water, ice, and rocks) and the overlying atmosphere. Horizontal heat conduction and transport in soil can be safely neglected. For this reason, the land surface models ignore horizontal interactions and are concerned especially with vertical interactions (Pitman 2003). Currently, sophisticated representations of these processes are included in the state-of-the-art general climate models (Ménard et al. 2015). The land models are mainly used to solve (at each time step) the following balances: surface energy balance; surface water balance (in some cases surface snow balance); and carbon balance (plant and soil carbon pools):

- Surface energy balance:

  This balance (Equation 8.8), which is considered as a perfect representation of the soil–atmosphere boundary condition, is mainly applied to determine the temperature of a soil horizon:

  $$R_n + H + LE + G = 0 \quad (8.8)$$

  where R_n: net radiation, i.e. is the difference between total incoming and outgoing radiation (W m^{-2}); H: convective heat transfer between the soil surface layer and the atmosphere (W m^{-2}); LE: latent heat due to evapotranspiration (W m^{-2}); and G: heat transfer from and to the ground (W m^{-2}). The equations used to calculate these different parameters (i.e. R_n, H, LE, and G) were reported with details in Chalhoub et al. (2017).

- Surface water balance:

  This balance (Equation 8.9) is mainly applied to describe the water flow in and out of a system. The general surface water balance is determined as follows:

  $$P = E_S + E_T + E_C + R_{Surf} + R_{Sub-Surf} + \left(\frac{\Delta SM}{\Delta t}\right) \quad (8.9)$$

  where P: is rainfall (mm); E_S: is soil evaporation (mm); E_T: is transpiration (mm), E_C: is canopy evaporation (mm); R_{Surf}: is surface runoff (mm), $R_{Sub-Surf}$: is sub-surface runoff (mm); and $\Delta SM/\Delta t$ is the change in soil moisture over a time step (mm). Details about these parameters were reported in Tsutsumi et al. (2004).

- Carbon balance:

  This balance (equations 8.10 to 8.12) is mainly applied to estimate the difference between carbon dioxide uptake by ecosystems (photosynthesis) and carbon dioxide loss to the atmosphere by respiration. The carbon balance is determined as follows:

  $$NPP = GPP - R_a = (\Delta C_f + \Delta C_s + \Delta C_r)/\Delta t \quad (8.10)$$

  $$NEP = NPP - Rh \quad (8.11)$$

  $$NBP = NEP - Combustion \quad (8.12)$$

where NPP: is net primary production; GPP: is gross primary production; Ra: is autotrophic (plant) respiration; Rh: is heterotrophic (soil) respiration; ΔCf, ΔCs, ΔCr: are foliage, stem, and root carbon pools; NEP: is net ecosystem production; NBP: is net biome production; and combustion: is carbon loss during fire.

8.2.4 Sea Ice Model

This computer model is mainly designed to simulate the development, melt, and movement of sea ice (see Figure 8.3).

The sea ice model has been integrated into several general climate models and ocean models (Horvat et al. 2017). This integration leads to very good estimations of several properties of sea ice. The large-scale dynamic of sea ice is simulated in the model as a two-dimensional continuum (Tonboe et al. 2016). This dynamic is mathematically represented using Newton's second law for sea ice (called also the first keystone equation):

$$m\frac{du}{dt} = -mfk \times u + \tau_a + \tau_w - mg\nabla\mu + \nabla.\sigma \qquad (8.13)$$

where m: is the mass per unit area of saline ice on the sea surface; u: is the drift velocity of the ice; f: is the Coriolis parameter; K: is the upward unit vector normal to the sea surface; τ_a and τ_w: are the wind and water stress on the ice, respectively; g: is acceleration due to gravity; μ: is sea surface height; and σ: is the two-dimensional stress tensor within the ice (Hunke and Dukowicz 1997). Ice thickness,

FIGURE 8.3 Development, melt, and movement of sea ice.

ice roughness, and ice concentration are fundamental data for calculating all these terms (i.e. m, u, f, K, τ_a, τ_w, μ, and σ) (Holland et al. 2012).

8.3 NUMERICAL RESOLUTION OF THE BASIC EQUATIONS

The different equations used to model the processes of climate systems are Partial Differential Equations (PDEs). For solving these equations, several methods have been developed. The common methods include especially the finite difference method (FDM) and the finite element method (FEM) (Smith 1978; Hughes 1987). Given a PDE with well-known terms and well-specified initial and boundary conditions, FDM and FEM algorithms compute the close value of the solution at a discrete number of points, providing a look-up table that can be interpolated when the solution is required elsewhere in the domain (Press et al. 1986; Sanchirico and Wilen 2005):

- Finite difference method (FDM):

    The basic principle of this method is to replace the derivatives of an unknown function by the difference quotients of unknown functions. The main objective of FDM is to replace a continuous field problem with infinite degrees of freedom by a discretized field with finite regular nodes. Accordingly, the domain is partitioned in space and in time and approximations of the solution are computed at the space or time points. The small discrepancy between the obtained solution with this method and the exact solution (or "very correct" solution) is determined by the error which is committed when replacing the derivative by the differential quotient. The derivative of u(x) with respect to x is mathematically expressed as follows (Thomas 1995):

    $$\begin{aligned} u'(x_i) = u_x(x_i) &= \lim_{\Delta x \to 0} \frac{u(x_i + \Delta x) - u(x_i)}{\Delta x} \\ &= \lim_{\Delta x \to 0} \frac{u(x_i) - u(x_i - \Delta x)}{\Delta x} \\ &= \lim_{\Delta x \to 0} \frac{u(x_i + \Delta x) - u(x_i - \Delta x)}{2\Delta x} \end{aligned} \qquad (8.14)$$

    A perfect approximation is highly ensured when the error committed in the FDM tends towards zero when Δx tends to zero (Cheney and Kincaid 1999). If the function u is sufficiently smooth in the neighborhood of x, it is so possible to quantify this error based on a Taylor expansion (Equation 8.15):

    $$u(x_i + \Delta x) = u(x_i) + \Delta x u'(x_i) + \frac{\Delta x^2}{2} u''(x + \Delta x_i) \qquad (8.15)$$

where Δx_i is a number from 0 to Δx (i.e. $x + \Delta x_i$ is point of the interval $[x, x + \Delta x]$). For the treatment of problems, it is so practical to retain only the first two terms of the previous expression as follows:

$$u(x_i + \Delta x) = u(x_i) + \Delta x u'(x_i) + O(\Delta x^2) \qquad (8.16)$$

where the term $O(\Delta x^2)$ shows that the error of the approximation is proportional to Δx^2. From Equation 8.15, we can deduce that there exists a constant C (higher than 0), such that for $\Delta x > 0$ sufficiently small we get the following equation:

$$\left| \frac{u(x_i + \Delta x) - u(x_i)}{\Delta x} - u'(x_i) \right| \leq C\Delta x, C = \sup_{y \in [x, x+\Delta x]} \frac{|u''(x_i)|}{2} \qquad (8.17)$$

Finally, it is important to mention that finite difference method has some limitations, e.g. the non-smoothness of the function which describes the dependence of the equilibrium point on the speed (Hoffman and Frankel 2001).

- Finite element method (FEM):

The FEM approximates the unknown function over the domain of interest (Chaskalovic 2008). To resolve a given problem, three steps are performed. These steps represent the principle of FEM. In a first step, the large system is subdivided into smaller and simpler parts (called finite elements). Here, the word "finite" reflects the limited number of degrees of freedom used to simulate the behavior of each element. The elements are assumed to be connected to one another, however, only at interconnected joints, known as nodes. In a second step, the algebraic equations that simulate these finite elements are then assembled into a larger system of equations that simulates the entire problem. Finally, the method uses variational methods from the calculus of variations to approximate a solution by minimizing an associated error function. Many methods have been developed to transform the physical formulation of a given problem to its finite element discrete analogue. The most common method is the Galerkin method (Bramble and Xu 1989). The basic principle of this powerful numerical method can be shown on the following one-dimensional linear parabolic partial differential problem (Hairer and Wanner 2010):

$$u_t(x, t) - (a(x)u_x(x, t))_x + b(x)u(x, t)) = f(xt)), x \in (0, 1), 0 \leq t \leq T$$
$$u(x, 0) = g(x), x \in (0, 1),$$
$$u(0, t) = u(1, t) = 0, 0 \leq t \leq T.$$

$$(8.18)$$

where $a(x) > 0$ and $b(x) \geq 0$ on $[0,1]$.

A spatial simple formulation of the given problem is determined as the product between the equation 8.18 and a test function v in the Sobolev space H_0^1, integrating over (0,1) and integrating by parts to show that $u \in H_0^1$ responds to the following condition:

$$(v, u_t(x,t)) + A(v, u(x,t)) = (v, f(x,t)), v \in H_0^1, 0 \le t \le T \\ u(x,0) = g(x), x \in (0,1), \tag{8.19}$$

Where

$$(v, u(x,t)) = \int v(x) u(x,t) dx, A(v, u(x,t)) \\ = \int \left[a(x) \frac{dv(x)}{dx} u(x,t) + b(x) v(x) u(x,t) dx \right] \tag{8.20}$$

On the other hand, the semi-discrete formulation of the Galerkin method refers in a first step to the subdivision of the domain of interest (0, 1) into many elements (n elements, $\mathbf{I}_i = (xi, xi+1)$, $i = 0,1,\cdots,n$) and then to the construction of a finite-dimensional subspace $S_p^n \subset H_0^1$ (Karashian and Makridakis 1998). S_p^n refers to piecewise p-degree polynomial functions spanned by a set of basis functions ϕ_i, $i = 1,2,\cdots, N$. For instance, $N = n-1$, for piecewise linear finite elements. Finally, the semi-discrete Galerkin formulation consists of finding:

$$u_N(x,t) = \sum_{i=1}^{N} c_i(t) \varphi_i(x) \tag{8.21}$$

where coefficients are time-dependent.

Similar to FDM, FEM has also some disadvantages (Aziz and Kellogg 1981; Babuška and Guo 1996). The major disadvantages are: (i) large amount of data is needed as input for the mesh used in terms of nodal connectivity and other parameters; (ii) the method needs a digital computer and fairly extensive; (iii) the method needs longer execution time compared with FEM (Cohen and Fauqueux 2000; Düster et al. 2001).

8.4 INPUT AND OUTPUT DATA

Numerical climate modeling requires several input data, usually derived from direct observations and previous modeling studies (Collins et al. 2011). The major input data are:

- Earth's characteristics (such as earth's radius, rotation period, land topography, coastline, ocean depth, soil characteristics, etc.);

- Boundary conditions for all sub-systems not clearly integrated in the climate model (such as distribution of vegetation cover, topography of ice sheets, etc.);
- Radiative forcing (such as solar energy intensity, energy perturbation, annual solar irradiance, etc.);
- Global greenhouse gases emissions (especially annual gridded carbon dioxide, methane, and nitrous oxide emissions). The climate model keeps track of radiation in every atmospheric grid box and at every time step. This is the primary way in which a change in greenhouse gases will affect the climate of the model;
- Land use and management (example: carbon dioxide emissions from land cover changes);
- Ozone (time-evolving 3D concentrations for forcing in models that do not include interactive chemistry);
- Volcanic forcing (example: dust and sulfur emissions).

Also, climate modeling provides much output data, the interpretation of which drives forward climate science, helping scientists evaluate the effects of human activities on the Earth's climate (Neale et al. 2013). Some output data are summarized in Table 8.2. The final step of climate modeling is the validation of model outputs.

Despite the perfect design of climate models, it is important to mention that there is no guarantee that these models will be adequate for their intended use: some processes treated as negligible can turn out to be more important than initially thought (Nolan et al. 2017). Consequently, climate models have to be tested and validated to assess their quality and evaluate their performance (Dee et al. 2011). Indeed, the simulation results of each climate model should be very close to the observations. The results of climate models are usually validated under climate conditions of a given period

TABLE 8.2
Some Outputs of Climate Models

Output	Unit
Precipitation rate	$kg\ m^{-2}\ s^{-1}$
Surface temperature	°C
Sea level rising	m
Sea level pressure	Pa
Thickness snow	m
Surface roughness length	m
Surface specific humidity	$kg\ kg^{-1}$
Relative humidity	%
Wind speed	$m\ s^{-1}$

(i.e. reference period) based on a comparison between data from direct observations and simulated data. Several statistical indicators are used to perform this comparison. Among these statistical indicators, we can cite: root mean square error, RMSE (Eq. 8.22), coefficient of model efficiency, CME (Eq. 8.23), and relative error RE (Eq. 8.24).

$$RMSE = \sqrt{\frac{1}{n}\sum (S_i - M_i)^2} \tag{8.22}$$

$$CME = \frac{\sum_{i=1}^{n}(M_i - \overline{M})^2 - \sum_{i=1}^{n}(S_i - \overline{M})^2}{\sum_{i=1}^{n}(M_i - \overline{M})^2} \tag{8.23}$$

$$RE(\%) = \frac{1}{n}\sum_{i=1}^{n}\frac{S_i - M_i}{M_i} \times 100 \tag{8.24}$$

where M_i is the observed data at time i, S_i is the simulated data at the same time, $\overline{M}$ is the average of observed data, and n is the observations number. Using these statistical indicators, the range of different future model simulations is taken to represent both the response of the climate system to forcing and the imperfect representation of it in the climate models (Ylhaisi et al. 2010).

8.5 STRENGTHS AND LIMITATIONS OF CLIMATE MODELING

Firstly, climate modeling is a highly effective tool to simulate quantitatively the different interactions of the major drivers of climate system (i.e. atmosphere, oceans, land surface, and ice). These quantitative simulations are often exploited for various purposes. The major purpose is the projections of future climate and the development of suitable adaptation strategies against climate change effects (Bollmeyer et al. 2015). Climate models can explain the main trends of past climate change (past 250 years). Having acknowledged our past climate system, climate models are extensively applied for future projections using various emission scenarios, giving ranges of future climate change as summarized in the reports of the Intergovernmental Panel on Climate Change (e.g. IPCC 2013). Despite the advantages of climate modeling, the complexity of our Earth imposes important limitations on existing climate models. The major disadvantages of climate modeling include the following points: longer-term forecasting because as models go further into the future they tend to diverge from each other (Wilks 1992); difficulty in analyzing general climate models results due to the associated models complexity; large uncertainties in aerosol forcing and ocean heat uptake; prohibitively expensive computational power required for uncertainty analysis using complex models; too coarse resolutions of general climate models for regional climate impact assessments (Weigel et al. 2010).

8.6 CONCLUSIONS

In this chapter, we have described the main components of climate models, analyzed the numerical resolution of basic equations, discussed input and output data of these models, and have summarized the major strengths and limitations of climate modeling. Preliminary analysis shows that climate modeling is an effective tool for projecting future climate conditions and for developing suitable adaptation strategies against climate change effects. Likewise, as climate component models (e.g. atmosphere, land ice, sea ice, and ocean) become more intimately coupled, perfect coupling options must be developed, and sea-ice components will need to be fully validated within the coupled context. Fresh numerical methodologies and algorithm improvements play a key role in the development process, as climate models continue to push the limits of computational power.

9 Coastal Water Resources and Climate Change Effects

9.1 INTRODUCTION

Coastal regions (Figure 9.1) are commonly defined as the intersection areas between the land and the sea (Clark 1996). This intersection leads to an environment with a distinct structure, diversity, and energy flow. This dynamic intersection leads also to a continuous change in coastal zone properties (Islam et al. 2009). The coastal regions include salt marshes, mangroves, wetlands, estuaries, and bays and are home to more than 50% of the global population. Globally, about 1.2 billion people live in the coastal regions (Burke et al. 2001). Furthermore, the coastal regions are potentially rich in natural soil and water resources with ecological, economical, and social significance (Fontaine et al. 2015).

The coastal regions are used extensively and increasingly for various human activities (e.g. human settlement, agriculture, trade, industry, and amenity). These activities are not always compatible and may lead to negative effects on soil and water resources, especially in the coastal agricultural areas. In addition to the human activities, climate change is negatively impacting coastal regions (Gillanders et al. 2011; IPCC 2013). Climate change can affect the coastal agricultural areas through direct and indirect ways: directly through sea level rise, storm surges, floods, and droughts and indirectly through events such as river floods, pulses, and quality of runoff that originates off-site but whose consequences affect the coasts (Wernberg et al. 2013). Also, climate change may pose the greatest threat to agriculture by increasing the demand for water and available water supply, and significant degradation of water quality (IPCC 2013).

Because climate change may result in significant alteration of natural water resources, as well as negative consequences on various agricultural activities in shoreline space, its impacts must be understood and countered. Consequently, this chapter describes available water resources in the coastal regions and then discusses the effects of climate change on these resources.

9.2 COASTAL WATER RESOURCES

Water resources are one of the most controlling factors of agricultural development in coastal regions (Adelana et al. 2008). The demand for water (especially for irrigation purposes) is growing in several coastal regions, especially crucial in regions that do not receive important amounts of rainfall such as the

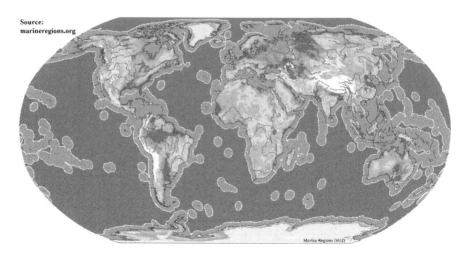

FIGURE 9.1 World map of coastal regions.

southernmost parts of Europe and the countries lying along the African coast and in the eastern Mediterranean (Attandoh et al. 2013). Several water resources are available in coastal regions (e.g. surface resources, shallow groundwaters, deep groundwaters, etc.). However, groundwater contained in coastal aquifers is considered as the most important water resource. Accordingly, in the present chapter, we focus on coastal aquifers. Worldwide, these aquifers are vital freshwater resources. They provide water for many important uses such as domestic water supplies, irrigation of crops and pastures, and industrial processes. Coastal aquifers also provide base flow to creeks and rivers during dry periods, thus supporting diverse ecosystems (Adelana et al. 2010).

9.3 CLASSIFICATION OF COASTAL AQUIFERS

A coastal aquifer is a wet underground layer of water-bearing permeable rock or unconsolidated materials from which groundwater can be usefully extracted using a spring installed in a coastal area (Salako Adebayo and Adepelumi Abraham 2018).

Generally, coastal aquifers are classed into two major groups: confined aquifers and unconfined aquifers (Figure 9.2).

9.3.1 Confined Aquifers

Confined aquifers are sandwiched between confining rock layers (i.e. geologic unit with little or no intrinsic permeability, for example clay materials). Due to these confining rock layers, groundwater in the confined aquifers is usually under high pressure. This leads to the rising of the water table in the well to a level *higher than* the water level at the top of the aquifer. Pressure surface is

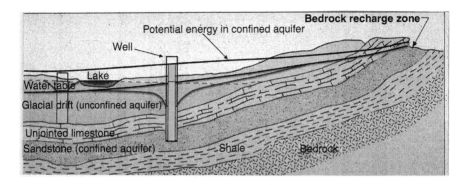

FIGURE 9.2 Confined aquifers and unconfined aquifers.

used to refer to the water level in the well (Zodhy et al. 1974). Groundwater flow through the confined aquifers is vertical and horizontal. The flow rates depend mainly on gravity and geological formations in the coastal areas (Fetter 2001). In addition to its key role (serving to hamper the movement of water into and out of the aquifer), the confining rock layers also serve as a barrier to the flow of contaminants from overlying unconfined aquifers. Contamination of confined aquifers can occur through a poorly constructed well or through natural seepage (Tadesse et al. 2010). The confined aquifer is actually unconfined at the recharge zone, the zone where water is able to seep into the ground and refill the aquifer. In order for pressure to build, the water level in this zone must be at a higher elevation than the base of the confining layer. When a well is drilled through the confining layer, often far from the recharge zone, the water in this well will rise to the level of the water at the recharge zone. In some instances this may be above the surface of the ground, in which case the well is called an artesian well (Toens et al. 1998).

9.3.2 Unconfined Aquifers

They refer to the aquifers into which water seeps from the ground surface directly above the aquifer. They are usually located near the land surface and have no confining rock layers (i.e. clay layers). The water table lies relatively above impermeable clay rock layers (Fetter 2001). Due to its localization (i.e. close to the land surface), easy groundwater infiltration by land pollutants leads usually to more vulnerability of unconfined aquifers to pollution compared to the confined aquifer (Freeze and Cherry 1979). Groundwater fluctuation follows with a certain lag, depending on the depth and nature of the unsaturated zone, the variation in precipitation, and the stored up groundwater in the space of the aquifer which in turn influences the rise or fall of water levels in wells that derive their source from unconfined aquifers (Hyndman and Gorelick 1996). The unconfined aquifer is called a "perched aquifer" if the following situation is ensured: groundwater bodies are separated from their main groundwater source by relatively impermeable rock layers of small areal extents and zones of aeration above the main body of groundwater. This perched aquifer is often a relatively small body of water and is only recharged by local rainfall (Winter et al. 1998).

9.4 PROPERTIES OF COASTAL AQUIFERS

Human activities (e.g. over-removal of groundwater), as well as extreme climate events (e.g. drought and sea level rise) often lead to saltwater intrusion with significant effects on water quality, quantity, and demand, which globally threatens coastal water security. Hundreds of millions of people could be directly influenced by this saltwater intrusion (Silliman et al. 2010; Shao et al. 2013). Thus, understanding the key properties (e.g. hydrogeology) of coastal aquifers is critical for groundwater development and management planning in coastal regions (Falgàs et al. 2011). Limited knowledge of these properties often leads to misleading decisions with regards to management of aquifers, resulting in improper knowledge and measures about pollution, overexploitation and seawater intrusion in coastal aquifers (Taylor and Greene 2008). The major properties that help in characterizing coastal aquifers are the following.

9.4.1 Porosity

It is the primary aquifer property that controls water storage. Porosity of a coastal aquifer can be defined as the percentage of the geological formation hosting an aquifer not occupied by solids. The higher the porosity, the more groundwater an aquifer can store (Niwas and Celik 2012); in other words, the proportion of solids to voids in a rock or soil sample formation. Accordingly, the mathematical representation of porosity (n) is given by the following formula:

$$n = \frac{V_p}{V_t} \qquad (9.1)$$

where V_p is the pores of rock or soil sample and V_t is the total volume of pores and solid material. If there is not a connection between the pores of rock (e.g. glassy volcanic rock), only a certain fraction of these pores allow for water movement. The effective porosity is mainly used as a term to refer to this fraction. The porosity of a coastal aquifer depends on several factors such as the degree of sorting (i.e. grains arrangement) and the angularity of grains. Well-sorted samples usually have higher porosity than poorly sorted samples. Poorly sorted samples contain grains that tend to fill in the void spaces. Also, samples with angular grains tend to have lower porosity than well-rounded grains, especially of similar sizes. The porosity of rocks and unconsolidated sediments varies considerably. In Table 9.1, some examples are summarized.

The porosity typically significantly affects the hydraulic properties of the aquifer such as groundwater flow (Buxton and Modica 1992). Because

TABLE 9.1
Some Examples of Porosity of Rocks and Unconsolidated Sediments

Rocks/Unconsolidated Sediments	Porosity (%)
Fractured basalt (rock)	0.05–0.50
Karst limestone (rock)	0.05–0.50
Sandstone (rock)	0.05–0.30
Limestone, dolomite (rock)	0.00–0.20
Shale (rock)	0.00–0.10
Fractured crystalline (rock)	0.00–0.10
Dense crystalline (rock)	0.00–0.05
Fractured basalt (rock)	0.05–0.50
Karst limestone (rock)	0.05–0.50
Gravel (unconsolidated sediments)	0.25–0.40
Sand (unconsolidated sediments)	0.25–0.50
Silt (unconsolidated sediments)	0.35–0.50
Clay (unconsolidated sediments)	0.40–0.70

groundwater flow significantly affects many geological and environmental problems (Bredehoeft and Norton 1990), the porosity of an aquifer should be highly considered during groundwater development and management planning in the coastal regions (Buxton et al. 1991). On the other hand, it is important to note that under low porosity conditions (usually n<5%), groundwater recharge can sometimes be significant, as in the case of fractured aquifers. These aquifers are an important and widely used class of aquifer, because they are commonly both highly permeable and rapidly recharged. For example, groundwater recharge to the limestone aquifer beneath Nittany Valley in the Spring Creek watershed (USA) is around 30–45% of the annual rainfall (in comparison to typical recharge of < 10% of rainfall). Fractured aquifers are permeable despite their overall low porosity (usually <5%) because natural fractures form in consistent orientations and are well connected in networks over hundreds of kilometers.

9.4.2 Hydraulic Head

The hydraulic head is the elevation to which water will naturally rise in a well. It may be measured as the depth below the natural land surface or against sea level to compare between different bores. Groundwater will always move from high to low hydraulic head. Mathematically, the hydraulic head is based on Darcy's equation (Equation 9.2). In this latter, the flow rate through an aquifer is proportional to the cross-sectional area perpendicular to flow and is also proportional to the head loss per unit length in the direction of flow:

$$Q = K \frac{h_1 - h_2}{L} A \qquad (9.2)$$

where Q: is the flow rate of liquid through the aquifer (m^3 s^1); K: is the hydraulic conductivity (m s^1); A: is the cross-sectional area perpendicular to flow (m^2); and h_1 and h_2: are the hydraulic heads associated with points 1 and 2 (m).

9.4.3 Hydraulic Gradient

The hydraulic gradient (*i*) between points A and B is the slope of the hydraulic head between those points (Equation 9.3). It is always expressed as a fraction (e.g. 0.2). It is the driving force that causes groundwater to move in the direction of maximum decreasing total head. Groundwater levels determined from a number of observation wells spaced out over an area can be used to characterize the hydraulic gradient in the two horizontal directions (Raghunath 2002):

$$i = \frac{\Delta h}{L} = \frac{h_1 - h_2}{L} \qquad (9.3)$$

where Δh (=h_1–h_2): is the hydraulic head difference (m); L: is the distance between points A and B (m); and h_1 and h_2: are the hydraulic heads associated with points 1 and 2 (m).

9.4.4 Hydraulic Conductivity

Hydraulic conductivity, symbolically represented as K, is a measure of capacity of a geological formation to transmit water (Domenico and Schwartz 1990). Mathematically, it is defined as the rate of flow under a unit hydraulic gradient through a unit cross-sectional area of aquifer:

$$K = K_i \frac{\rho g}{\mu} \qquad (9.4)$$

where K: is the hydraulic conductivity (m s^{-1}), K_i: is the intrinsic permeability (m s^{-1}), ρ is the density of water, μ is the dynamic viscosity of water. Typical values of K for different materials vary from hundreds of m day^{-1} for gravels, to imperceptible rates (e.g. 0.5 m day^{-1}) for sandstone (Table 9.2). It is important to note that hydraulic conductivity, which is a function of water viscosity and density, is in a strict sense a function of water temperature; but, given the small range of temperature variation encountered in most aquifer systems, the temperature dependence of hydraulic conductivity is usually neglected (Tsang and Tsang 1987). Furthermore, it is important to note that variation in hydraulic conductivity values of rocks and soils is dependent on several factors such as weathering, fracturing, solution channels, and depth of burial (Zhu et al. 1997).

TABLE 9.2
Typical Hydraulic Conductivity of Geological Units
(Adapted from Domenico and Schwartz 1990)

Geological Unit	Hydraulic Conductivity (m d^{-1})
Fine sand	0.02 to 17
Coarse sand	0.08 to 520
Shale	8 x 10^{-9} to 2 x 10^{-4}
Gravel	26 to 2,592
Sandstone	3 x 10^{-5} to 0.5
Permeable basalt	0.03 to 1,728

9.4.5 Aquifer Transmissivity

Transmissivity, symbolically represented as T, is a measure of capacity of a geological formation to transmit groundwater throughout its entire saturated thickness. Accordingly, the mathematical representation of this aquifer property can be determined by the following equation:

$$T = K \times b \quad (9.5)$$

where T: is the transmissivity (m^2 s^{-1}), K: is the hydraulic conductivity (m s^{-1}), b: is the saturated thickness of the aquifer (m). Usually, the transmissivity of an aquifer is determined through a pumping test using the levels of drawdown over time pumped (McKinney 2015). However, this test is an intensive and costly activity and it is often only carried out to resolve a specific issue. Alternatively, the transmissivity can be determined using geophysical techniques such as electrical resistivity and seismic techniques (Kelly 1977; Huntley 1986). Good identification of the variation in transmissivity throughout an aquifer is so useful for identifying boundaries where values will typically be lower than elsewhere (Dewandel et al. 2004). The potential of an aquifer for exploitation is based highly on the transmissivity values. Areas with high transmissivity values can be attributed to having a thick aquifer and good potential for exploitation. In Table 9.3, we present the classification proposed by Offodile (1983) for identifying the potential of an aquifer for exploitation.

9.4.6 Storage Coefficient

The storage coefficient, symbolically represented as S, is a measure of the volume of water released from storage with respect to the change in hydraulic head and surface area of the aquifer (Connected Waters Initiative 2013). Mathematically, S is calculated as the volume of water released from storage per unit decline in hydraulic head in the aquifer, per unit area of the aquifer (Morris and Johnson 1967):

TABLE 9.3
Aquifer Potential for Exploitation Based on Transmissivity Values (Adapted from Offodile 1983)

Transmissivity ($m^2\ d^{-1}$)	Aquifer Potential for Exploitation
> 500	High potential
50–500	Moderate potential
5–50	Low potential
0.5–5	Very low potential
< 0.5	Negligible potential

$$S = \frac{dV_w}{dh} \times \frac{1}{A} \qquad (9.6)$$

where V_w: is the volume of water released from storage (m^3), h: is the hydraulic head (m), and A: is the area of the aquifer (m^2). The values of S are highly dependent on the types of aquifer (i.e. confined or unconfined aquifer):

- Unconfined aquifers:

  For these aquifers, the predominant source of water is from gravity drainage as the aquifer material is dewatered during pumping (Morris and Johnson 1967). Under this condition, S value equals approximately the aquifer porosity. For unconfined aquifers, S values range usually from 0.01 to 0.30 (Connected Waters Initiative 2013). Unconfined aquifers can provide more water for a smaller change in head compared to confined aquifers;

- Confined aquifers:

  For these aquifers, the predominant source of water is from compression of the aquifer and expansion of the water when pumped (Morris and Johnson 1967). During pumping, the pressure is decreased in a confined aquifer; however, the aquifer is not dewatered. For confined aquifers, S values range usually from 1×10^{-5} to 1×10^{-3} (Connected Waters Initiative 2013).

9.5 ESTIMATION OF AQUIFER PROPERTIES

In coastal aquifer studies, major properties of interest include hydraulic conductivity (K), transmissivity (T), and storage coefficient (S) (Busari and Mutamba 2014). These properties are crucial in coastal aquifer control and management. Identification of these properties leads to a good quantitative estimation of the hydraulic response of the aquifer to recharge and pumping. Furthermore, these properties influence the environment by controlling infiltration, irrigation rate, and hence the water movement through the ground

(Or and Tuller 2003). Hence, in this section, we are interested in the estimation methods of these properties. Pumping test, electrical technique, and electromagnetic technique are the main methods that are usually performed to estimate site-specific values for K, T, and S of aquifers (Damiata and Lee 2006):

- Pumping test:

  This field test involves pumping water at a controlled rate from a test well and then observing the responses of the aquifer (i.e. water-level change (drawdown), and flow rate change) in one or more surrounding observation wells and optionally in the pumped well (control well) itself (Figure 9.3).

  Recommended minimum frequencies for measurements of these responses are summarized in Table 9.4. These response data are exploited to determine the hydraulic properties of aquifers (i.e. K, T, and S), evaluate well performance, and identify aquifer boundaries (Sterrett 2007). Usually, K, T, and S are estimated from a constant pumping test (i.e. maintain pumping at the control well at a constant rate) by fitting mathematical models (type curves) to responses of the aquifer (drawdown) based on a procedure known as curve matching (Clark 1977). Prior to the realization of a pumping test (i.e. a field experiment), several previous

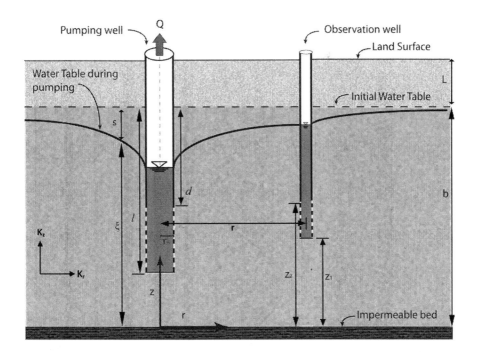

FIGURE 9.3 An overview of pumping test.

TABLE 9.4
Recommended Minimum Frequencies for Water Level Measurements for Pumping Test (Adapted from Kruseman and de Ridder 1990)

Minimum Frequencies for Water Level Measurements for Pumping Test:

In the Test Well	In the Observation Well
Every minute for the first 10 minutes	Every 10 minutes for the first 100 minutes
Every 2 minutes from 10 minutes to 20 minutes	Every 50 minutes from 100 minutes to 500 minutes
Every 5 minutes from 20 minutes to 50 minutes	Every 100 minutes from 500 minutes to 1000 minutes
Every 10 minutes from 50 minutes to 100 minutes	Every 500 minutes from 1000 minutes to 5000 minutes
Every 20 minutes from 100 minutes to 200 minutes	Every 24 hours from 5000 minutes onward
Every 50 minutes from 200 minutes to 500 minutes	Final water level measurement just prior to end of pumping
Every 100 minutes from 500 minutes to 1000 minutes	(–)
Every 200 minutes from 1000 minutes to 2000	(–)
Every 500 minutes from 2000 minutes to 5000 minutes	(–)
Every 24 hours from 5000 minutes onward	(–)

reconnaissance works should be taken into consideration such as the identification of a suitable time for pumping test, reconnaissance of aquifer units, identification of location and construction details of new and existing wells, and development of strategy for disposal of pumped water, etc. (Kruseman and de Ridder 1990).

Finally, it is important to note that to get the desired output data from the pumping data (i.e. obtain the values of K, T, and S for a given aquifer), it is necessary to use computer software programs (Kresic 1997). Several software programs were developed during recent decades. AquiferTest is one of these software programs. It is an easy-to-use software package for analyzing, interpreting, and visualizing pumping test data. It delivers all the tools needed to accurately interpret data from all types of aquifers in all types of test conditions (Rasmussen et al. 2003).

- Electrical technique:

The pumping test is an intensive and costly activity. Alternatively, a coastal aquifer could be investigated using electrical resistivity (ER) technique. ER provides an expensive investigation of a coastal aquifer with minimal equipment (Pawar et al. 2009). ER technique is based on the introduction of a time-varying direct current (DC) or very low frequency (<1 Hz) current into two grounded electrodes to generate potential differences as measured at the surface with units of Ohm-meters (Ω-m). The electrodes by which current is introduced into the ground are called current electrodes and electrodes between which the potential difference is measured are called potential electrodes (Figure 9.4).

A deviation from the norm in the pattern of potential differences expected from homogeneity proffers the necessary information on the form and electrical characteristics of subsurface inhomogeneities (Matula 1979). As current is been injected into the ground, corresponding potential difference (ΔV) is measured. Then, the measured ΔV, intensity of injected current (I), and electrode spacing (a) are used to determine electrical resistivity (ρ_a) based on Ohm's law (Lowrie 2007):

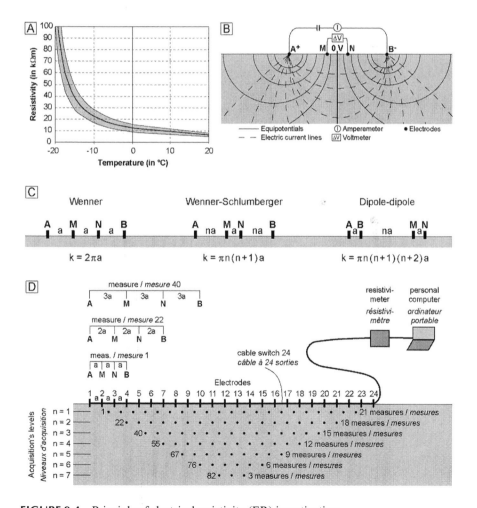

FIGURE 9.4 Principle of electrical resistivity (ER) investigation.

$$\rho_a = \frac{\Delta V}{I} \times 2\pi a \qquad (9.7)$$

The earth materials have a wide range of electrical resistivity values. In Table 9.5, we summarize the typical electrical resistivity values for some common earth materials.

As reported in several studies (e.g. Oseji et al. 2005; Omosuyi et al. 2007; Alabi et al. 2010), the electrical resistivity measurements have been shown to be able to infer the lithological information and hydrogeological parameters required for mapping aquifers. For aquifer mapping, electrical conductivity measurements (inverse of electrical resistivity) are used. In this context, the interest is the delineation of connected pore spaces, void spaces, interstices, fractures within rocks that are water filled which allows for decreased resistivity values and increased conductivity values. However, more information is still required as high conductivity within rock formation or units could be due to a number of things besides water, some of which include the presence of clay minerals, contamination plumes, etc. (Loke and Barker 1996).

- Electromagnetic technique:

Currently, the electromagnetic inductive technique (EM) is widely used to investigate geological formations and aquifers by mapping variations in the electrical conductivity of the ground. It is a powerful technique for getting information about electrical ground conductivities (Das 1995). EM is widely applied for mapping industrial pollution in groundwater, mapping groundwater quality, and mapping soil salinity in connection with the growth of crops. This technique measures the response of the

TABLE 9.5
Typical Electrical Resistivity Values for Some Common Earth Materials (Adapted from Hoekstra and Blohm 1990)

Element	Ionic Form
Sand (dry)	10^3–10^7
Sand (saturated)	10^2–10^4
Clay (dry)	3700
Clay (saturated)	20
Basalt (saturated)	100
Limestone (saturated)	40
Silts	10^2–10^3
Shales	10^1–10^3
Granite	10^3–10^5
Coal	10^4

ground to the propagation of incident alternating electromagnetic waves which are made up of two orthogonal vector components (an electric intensity (E) and a magnetizing force (H)), in a plane perpendicular to the direction of travel as shown in Figure 9.5 (Beck 1981). First of all, an alternating electric current (AC) is applied to a transmitter coil. Then, immediately after this current application, a primary electromagnetic field in the coil is generated. This induces small electric currents in the ground, generating a secondary magnetic field, called a ground magnetic field which depends mainly on coil spacing, operating frequency, and ground conductivity (Lashkarripour 2003). The frequencies of electromagnetic radiation range from atmospheric micropulsations which have frequencies less than 10 Hz, through to the radar bands of frequencies between 108 and 1011 Hz up to X-rays and gamma-rays of frequencies far beyond 1016 Hz (Reynolds 1997).

The value of apparent conductivity can be estimated through direct interpretation using multiple electromagnetic readings at selected locations (based on empirical formulae) or often using EMIX34 computer software. The value of apparent conductivity depends on several factors. The major factors include soil structure, soil porosity, clay content (if present), temperature, and degree of water-saturation in the soil (Hoekstra and Blohm 1990). EM techniques are divided into two major types: frequency EM and time EM. In the first type, high or low frequency transmitters are used. The high transmitter frequencies lead to high-resolution investigation of conductors (shallow depths), whereas the high transmitter frequencies lead to higher depths of investigation. In the second type, secondary magnetic field is measured

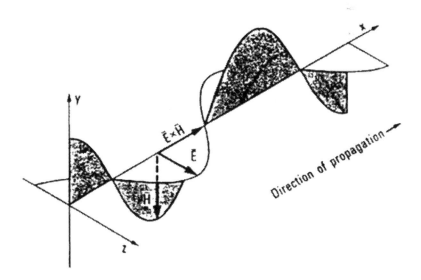

FIGURE 9.5 Principle of electromagnetic inductive technique (adapted from Beck 1981.)

as a function of time, with early-time measurement being suited best for shallow depths while late-time measurement yields results of the higher depths. It is paramount to note that depth of investigation is also being governed by coil configuration; measurements from coil separations are affected by electrical properties, thus the larger coil separation investigates higher depths, while the smaller coil separation investigates shallow depths.

9.6 EFFECTS OF CLIMATE CHANGE ON COASTAL AQUIFERS

Climate change (i.e. more extreme weather events) can lead to longer periods of drought and floods, which have direct effects on the availability and dependency of coastal aquifers. In long periods of drought, there is a higher risk of depletion of coastal aquifers (especially shallow aquifers). People in water-scarce coastal areas will increasingly depend on coastal aquifers, due to their buffer capacity (Allen et al. 2004). Furthermore, indirect climate change effects, especially intensification of human activities and land use modifications are leading to a significantly increased demand for coastal aquifers throughout the world. The strategic use of coastal aquifers for global water and food security under climate change conditions is becoming more important. Accordingly, coastal aquifers should have a more prominent role in climate debates (Ali et al. 2012). Climate change affects water quantity in coastal aquifers through an important change in aquifer recharge and also water quality through sea level rise which can lead to saltwater intrusion into coastal aquifers. Assessing the effects of climate change on groundwater resources in coastal regions requires focusing on changes in groundwater recharge and sea level rise, on the loss of fresh groundwater resources in water resources-stressed coastal aquifers (Chen et al. 2004). Accordingly, in the following sections, we discuss in detail the effects of climate change on aquifer recharge and then how sea level rise can lead to saltwater intrusion:

9.6.1 Effects of Climate Change on Aquifer Recharge

Climate change can significantly alter aquifer recharge rates for major aquifer systems and thus affect the availability of fresh groundwater. Climate change is likely to affect aquifer recharge rates due to changes in rainfall, temperature and evapotranspiration (Cave et al. 2003). Increased temperature under a changing climate usually leads to a substantial increase in evapotranspiration rate which in turn leads to significant decreases in runoff and recharge (i.e. decreased surface flow with less water available for aquifer replenishment). Furthermore, as revealed by recent international studies (e.g. Cave et al. 2003; Green et al. 2011), sensitivity of recharge to changes in rainfall indicates that regions with high current recharge rates are less sensitive to change in rainfall. The significant sensitivity of recharge to changes in rainfall is observed in the arid and semi-arid regions (rainfall amount < 500 mm year^{-1}). For example, based on a simplistic empirical relationship between average annual rainfall and annual recharge rate, Cave et al.

(2003) reported that a 20% decrease in rainfall over the central parts of Southern Africa could lead to an 80% decrease in recharge rate for areas that currently receive lower than 500 mm rainfall per annum. On the other hand, based on modeling approaches, Döll 2009 and Pitz (2016) recommend downscaled regional climate models to project climate change scenarios of changes in climate variables, especially rainfall and temperature. These downscaled climate values are used as input data into hydrological models to determine the effects of climate change on the recharge of coastal aquifers (Dennis and Dennis 2013.). For example, Döll (2009), after the projection of climate change effects on aquifer recharge by 2050, reported that the spatial pattern and intensities of changes vary with the CO_2 emissions scenario and the climate model used to translate emissions into changes of climatic variables. However, all scenarios agree broadly that aquifer recharge will increase in northern latitudes but will substantially decrease by 30–70%, or even more than 70%, in many semi-arid regions such as the Mediterranean regions, northeastern Brazil and southwestern Africa. According to the results of a global hydrological model, aquifer recharge, when averaged globally, increases less than total runoff (by 2% as compared with 9%) until the 2050s for the global model ECHAM4 climate change response to the A2 scenario (i.e. highest CO_2 emission scenario).

In addition to climate change effects, it is important to note that aquifer recharge in the coastal regions highly depends on the properties of the aquifer media and the properties of the overlying soils (Bouraoui et al. 1999). Many methods can be used to evaluate recharge based on surface water, unsaturated zone, and aquifer data. Among these, numerical modeling is the only method that can estimate aquifer recharge (Brouyere et al. 2004). This method is also highly useful for estimating the relative importance of various factors that can have direct and indirect effects on aquifer recharge, thereby ensuring that the model realistically accounts for all the processes involved. However, the accuracy of model estimations (i.e. estimated aquifer recharge) greatly depends on the availability of reliable hydrogeological data and the collection of all required climate data (Holman 2006).

9.6.2 Sea Level Rise and Saltwater Intrusion in Coastal Aquifers

When considering water resources in coastal regions, coastal aquifers are key sources of freshwater. However, saltwater intrusion can be a serious problem in these regions. Saltwater intrusion refers to the replacement of freshwater in coastal aquifers by saltwater from the sea (Figure 9.6) which leads to a reduction of available fresh groundwater resources.

Saltwater intrusion occurs as a result of the landward movement of seawater into the coastal aquifer. This landward movement is caused by changes in the freshwater and saltwater pressure gradients, resulting mainly from three major factors:

- Over-pumping (i.e. over-exploitation) of the coastal aquifers (more freshwater leaving the aquifer results in decreased freshwater hydraulic head);

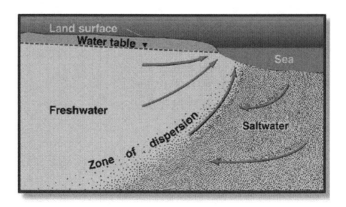

FIGURE 9.6 Saltwater intrusion into a coastal aquifer.

- Decreased recharge (less freshwater entering the aquifer results in decreased freshwater hydraulic head).
- Sea level rise (increased volume of saltwater results in increased saltwater hydraulic head). Sea level rise can lead to increased risk of intrusion and well pollution. Globally, mean sea level has been rising from 1.7 ± 0.5 mm year^{-1} for 1961–2003 to 3.1 ± 0.7 mm year^{-1} for 1993–2003. The IPCC (2007) predicted that by 2100 global sea levels will rise between 0.09 to 0.88 m, thereby affecting many vulnerable regions of the world.

Climate change typically leads to significant increases in the frequency and intensity of extreme climate events such as tropical storms and hurricanes (IPCC 2007). These increases coupled with sea level rise can lead to more frequent and intense storm surges to coastal areas, causing saltwater flooding in areas containing coastal wells (Van Biersel et al. 2007). With this flooding, wells would be subject to inundation and physical damage by storm surge. The duration of such inundation would likely be brief in the event of storm surge, and the risk of saltwater pollution would hence likely also be short-lived (IPCC 2013). Wells located in low-lying areas and other locations susceptible to storm surge will be particularly at high risk.

9.7 CONCLUSIONS

In the present chapter, we discussed in detail the key properties of coastal aquifers (i.e. porosity, hydraulic head, hydraulic gradient, hydraulic conductivity, transmissivity, and storage coefficient). Identification of these properties is critical for groundwater development and management planning in coastal regions. In addition, this chapter has highlighted the effects of climate change on the serious problems of aquifer recharge change, sea level rise, and saltwater intrusion in coastal regions. This chapter identifies the major factors that control these serious problems. Based on the finding of this chapter, we propose that the modeling approach could be a powerful tool for estimating effects of climate change on these problems. However, the accuracy of this approach depends highly on the availability of hydrogeological data and climate data and therefore, suitable methods for obtaining these data are required for the optimum evaluation of climate change effects on coastal aquifers.

10 Climate Change and Coastal Erosion

10.1 INTRODUCTION

Coastal regions are the most vulnerable to the erosion process (Fedorova et al. 2010). This global problem is usually destructive (i.e. "destructive coastal erosion") and occurs due to the combination of many factors, especially sea level rise, storm events, and storm surges that occur as a consequence of climate change (Ferreira et al. 2009). Destructive coastal erosion is already occurring in many regions of the world and has various negative impacts on farmland, coastal ecosystem, and rural infrastructure such as roads and sea walls (Mangor et al. 2017). These negative impacts are currently more serious now than they were 50 years ago (Łabuz and Kowalewska-Kalkowska 2011). Due to these impacts, coastal erosion has received wide attention (e.g. Vestergaard 1991; Tõnisson et al. 2008; Soomere et al. 2011). These studies have revealed that increased the erosion process will have considerable impacts on the sustainability of agriculture and human settlement in coastal regions which could hinder future food security (Florek et al. 2008). The recommendations of these studies include the need for a deeper knowledge of contributory factors to coastal erosion (especially the contribution of climate change) and the development of suitable strategies to protect coastal regions from the impacts of the erosion process. Accordingly, the present chapter describes the major contributory factors to coastal erosion and then discusses the potential adaption strategies to protect coastal regions from the impacts of erosion.

10.2 CONTRIBUTORY FACTORS TO COASTAL EROSION

Coastal erosion is a natural process, an endless sediment redistribution process that continually changes beaches, dunes, and bluffs. Waves, currents, tides, wind driven water, rainwater runoff, and groundwater seepages all move sand, sediment, and water along the coast (Mangor et al. 2017). Several factors can significantly increase coastal erosion. These factors can be subdivided into two types: natural factors (winds, storms, sea level rise, etc.), and human-induced factors (land reclamation projects, dredging activities) (Eurosion 2004). Knowledge on the contributory factors to coastal erosion and their interaction is essential for integrated coastal zone management. In the following sections, we review in detail these different factors.

10.2.1 Natural Factors and Coastal Erosion

The major natural factors include wind, waves, storms, near-shore processes, sea level rise, slope processes, and land compaction. Many of these natural factors occur because of climate change (Ferreira et al. 2009). Consequently, climate change directly and indirectly causes significant soil erosion in the coastal regions (Komar 2000):

- Winds:

    This natural factor significantly accelerates coastal erosion, making it more noticeable and problematic. Wind erosion (called also aeolian erosion) typically occurs as a consequence of wind removing soil from sandy coasts where the soil is not compacted or is of a finely granulated nature. This is highly visible along some sandy coasts of Aquitaine, Chatelaillon, Rosslare, and the Netherlands (Abbot and Halevy 2010). Aeolian erosion can result in several types of movement of the soil such as suspension, saltation, and creeping (Figure 10.1). Suspension occurs when the wind takes fine soil particles (fine dirt and dust) that are less than 0.1mm in diameter. Wind carries these particles over very long distances. With saltation movement, the wind briefly lifts up soil particles 0.1–0.5 mm in diameter and drops them in very short intervals. Creeping movement occurs when soil particles larger than 0.5 mm in diameter are dragged over the surface of the land because they are too heavy for the wind to lift (Alfaro 2008).

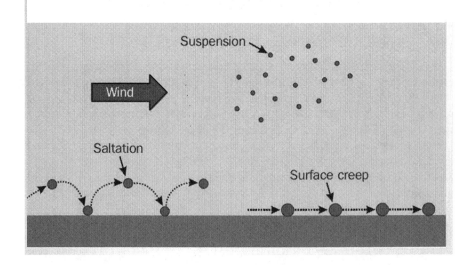

FIGURE 10.1 Types of movement of soil due to wind action.

- Waves:

    In this case, it is mainly the destructive waves that contribute to coastal erosion. These waves have an important energy and they are most powerful in stormy conditions. This is highly visible along open straight coasts such as those of Sussex, Ventnor, Aquitaine, Chatelaillon, Holland, Vagueira, Estella, Giardini Naxos, Ystad, Rostock, etc. Destructive waves have stronger impacts when their velocity increases. The wave velocity is proportional to the square root of he wave length (Holman 1986). This proportionality could be determined based on the following formula:

    $$V = \sqrt{\frac{gL}{2\pi}} \quad (10.1)$$

    where V: is the velocity of wave (m s^{-1}); L: is the length of wave (m); and g: is the acceleration of gravity (= 9.81 m s^{-2}). The coastal erosion due to destructive waves is a consequence of four major processes: hydraulic action, compression, abrasion, and attrition (Ruggiero 1997). The hydraulic action occurs when the motion of waves against the coastline produces mechanical weathering. Generally, hydraulic action refers to the ability of destructive waves to dislodge and carry away the materials of the coastline (soil or rock particles). The compression process refers to the compressive forces (air forces) that occur inside the rocks of the coastline (i.e. when the waves crash against the rock). Therefore, the compression process is visible along rocky coasts. The air inside the crack is rapidly compressed and decompressed causing cracks to spread and pieces of rock to break off.

    > The abrasion process occurs when rocks and other materials carried by the sea are picked up by destructive waves and thrown against the coastline, causing more material to be broken off and carried away by the sea. Finally, the attrition process occurs when material (especially rocks and stones) carried by destructive waves hit and knock against each other, thereby wearing them down (Kaminsky et al. 1997).

- Storms:

    Storms on the coastline are the result of a combination of three major factors: increased water level (called also storm surge), high speed of winds, and high action of the destructive waves along the coast (Almeida et al. 2012). The magnitude and destructive power of storms depends on the wind direction, wind speed, wave action, and the status of the tide at the time of the storm's passing (Figure 10.2).

    Coupled with high tides, storms have a high potential to cause considerable damage to coastal infrastructure, beaches, and dunes. For example, in a few hours, the storms have the potential to move back

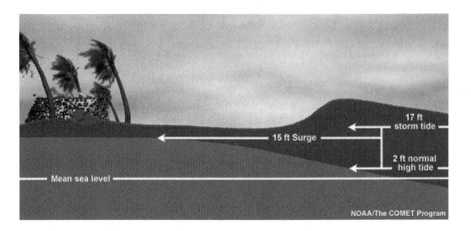

FIGURE 10.2 Contribution of storms to coastal erosion.

the dunes over 10 meters (Armaroli et al. 2012). Because of climate change, future storm events may produce stronger destructive waves, larger storm surges, and heavier rainfall, leading to higher flood effects and accelerated coastal erosion. Furthermore, climate change is anticipated to increase the frequency and intensity of storms (Zhang et al. 2000).

- Sea level rise:

  Climate change, especially global warming, has significantly increased the water temperature of seas and oceans. A significant rise in sea level has occurred over the past 50 years due to this increase in water temperature and also due to the increase in the amount of water trapped as ice on land (IPCC 2013). Generally, coastal erosion accelerates as sea level rises (DOE 1993; Huq et al. 1995). Based on the data reported in IPCC (2007), it is estimated that sea level rise will be between 18 and 59 cm higher in the decade 2090–2099 compared to 1980–1990. Several international studies (e.g. Hossain and Lin 2001; Feagin et al. 2005; FitzGerald et al. 2008), have concluded that sea level rise leads usually to significant coastal erosion and causes the shoreline to retreat landwards (Enríquez et al. 2017).

- Near-shore processes:

  These processes refer to the dispersion of water and sediments near the coast (Farhadzadeh et al. 2012). This dispersion can play a positive or negative role on sandy coasts. The positive role occurs when this dispersion leads to the formation of sandy beaches, whereas the negative role is observed when there is an erosion of sandy beaches (Gourlay 2010, 2011). Waves and the currents that they generate are the major factors in the transportation and deposition of near-shore sediments (Holthuijsen et al. 1989).

- Slope processes:

    These refer to the different land–sea interactions in coastal slopes. Water seepage in sloped landforms (e.g. bluffs) is an example of these interactions (Amin and Davidson-Arnott 1995). Water can fill in voids, thereby adding weight to the landform, causing the sediment to flow from its own accumulated weight (Anderson 2003). The flow of water throughout a landform can form seep zones (dark bands of moist soil) and springs, which may cause rill and gully erosion in the sloped landforms. Several factors can contribute to water seepage in sloped landforms such as natural precipitation, septic-system leaching, roof drains, and runoff from impervious surfaces such as paved driveways and parking lots (Brown et al. 2005).

10.2.2 Human-Induced Factors and Coastal Erosion

The major human-induced factors include land reclamation projects, river water regulation projects, land clearing activities, and mining activities.

- Land reclamation projects:

    Land reclamation projects refer to the creation of new lands in the sea that are carried out for different development activities such as agriculture, trade, industry, and tourism (Fang and Daniels 2005). During these projects, dredged sediments are used to build the new lands in the sea. Hence, these projects are known as dredging projects (Siham et al. 2008). These dredging projects have intensified in the past 20 years. An example of a major land reclamation project is the extension of the port of Rotterdam in the Netherlands with sand from the sea which was carried out in the 1970s (Merk and Notteboom 2013). Typically, these projects result in adverse impacts on coastal erosion (Apitz et al. 2005). Without proper management, dredging projects will cause negative effects on coastal erosion. When dredging projects are not well managed, geological hazards such as soil contamination, water contamination, and flooding may occur (Limeira et al. 2010). For example, about 80% of erosion on the east coast of Florida in the USA is attributable to the effects of unmanaged dredging projects (Groothuizen 2008).

- River water regulation projects:

    These projects can lead to an important decrease in the volume of sediment discharge of rivers. For example, from 1950 to 2004, the annual volume of sediment discharge of Ebro River (Spain) decreased by about 11%. In 1950 some water regulation works were built in and around the river (Eurosion 2004). This decreasing may be causing a high sediment deficit at the river mouth, and consequently erosion in the sediment cell (i.e. river damming). Furthermore, all activities that could lead to decreasing the water flow or preventing river flooding (as a major generator of sediments in the water system) are expected to decrease the volume of sediments reaching the coast (Kamphuis 2002). This is a good example of the effects of river water regulation works on coastal erosion (Gopinath and Seralathan 2005).

- Land clearing activities

  They refer to various activities that remove vegetation covering the land surface. Due to these activities, especially when they are extensive, land without any vegetation cover is exposed, leading to significant soil erosion (Young 1988). Furthermore, these activities can destroy an entire ecosystem, causing several environmental problems, such as greenhouse gas emissions, increased soil salinization, destruction of natural habitats for animals, and decrease or even extinction of indigenous flora and fauna (Jacobson 2005). Land-clearing activities occur worldwide and are especially prominent in the United States, Europe, Australia, and New Zealand (Eurosion 2004).

- Mining activities:

  They refer to various activities that aim to extract useful substances from the ground (e.g. coal mining, gold mining, etc.). The use of water jets during mining activities often mobilizes the upper soil layers, resulting in negative effects on the land and soil. Soil erosion is one of the major effects (Swenson et al. 2011). Deforestation and loss of fauna and flora are also other effects of mining activities. Furthermore, the mining activities may lead to climate change through an increase of greenhouse gas emissions (Batlle-Bayer 2010).

10.3 EROSION MODELING AS A TOOL FOR ASSESSING CLIMATE CHANGE IMPACTS

The effects of climate change on coastal erosion is complex (Ewen et al. 2000), and evaluating these effects through field research is difficult and expensive because numerous uncontrolled factors may affect coastal erosion in addition to the controlled factors. Also, the magnitude of coastal erosion in a short period (e.g. 1 month to 1 year) is usually small and may not be readily detectable by routine field research procedures (Lukey et al. 2000). Hence, modeling is the most effective tool to predict quantitatively the effects of climate change on coastal erosion. Several soil erosion models have been developed in the past years, each varying in terms of complexity, accuracy, input and output data, approaches, and their spatial and temporal scales (Kinnell 2005). These models have mainly been developed to represent the processes of detachment, transport, and deposition of eroded soil. Generally, these models can be subdivided into three categories (Jakeman and Hornberger 1993): empirical models (based mainly on statistical relationships between erosion process and key contributory factors), physically based models (in which various complex physical equations such as partial differential equations are used to represent erosion processes), and hybrid models (a mixture of dynamic and empirical soil is used to represent erosion processes).

10.3.1 Empirical Models

For the empirical erosion models, the Universal Soil Loss Equation (USLE) is the most commonly used model (Wischmeier and Smith 1965). In this case, the average annual soil erosion rate (E, t km^{-2} year^{-1}) is modeled as the product of six key factors: soil erosion constant (α, calibrated with published surface erosion rate data, average annual rainfall (R, mm year^{-1}) (mm), soil erodibility (K, depends mainly on soil texture), slope-length factor (L), slope-gradient factor (Z), and vegetation-cover factor (U):

$$E = \alpha R^2 K L Z U \quad (10.2)$$

The accuracy of this model depends highly on the quality of estimated factors (i.e. α, R, K, L, Z, and U). Therefore, these factors must be accurately estimated (Lufafa et al. 2003). Original values of such factor were derived from field measurements in various areas within the USA; however, they have been expanded and calibrated based on information gathered by researchers who have applied the USLE model in different countries in the world (Renard et al. 1997). Usually, the USLE model is used worldwide for predicting sediment losses from earthworks areas to support earthworks consent applications (Auckland Regional Council 2005). On the other hand, the USLE model could be used with projections of average annual rainfall increase to provide a good prediction of the impact of climate change on soil erosion.

10.3.2 Physically Based Models

These models (also called mechanistic or theoretical models) require several specific parameters describing various physical processes involved in soil erosion, including infiltration of rainwater into the soil, detachment of soil particles by raindrop splash, and surface runoff. All these processes are useful to accurately estimate the amount of soil loss by water. The erosion processes in the physically based models are mainly expressed through various complex physical equations such as partial differential equations that are based on several assumptions (Pandey et al. 2016). These equations use state variables. The physically based models can overcome several defects due to the use of parameters having physical interpretation (Abbott et al. 1986). Currently, together with GIS and remote sensing techniques, physically based models are widely used to study soil erosion. Over 50 physically based soil erosion and sediment yield models have been developed over the past years. These models were mainly developed to estimate soil detachment and sediment delivery processes at the hillslope scale (Devia et al. 2015). SWAT, WEPP, AGNPS, ANSWERS, and SHETRAN are the most common models used for soil erosion and sediment studies (Pandey et al. 2016).

10.3.3 HYBRID MODELS

Hybrid models represent models between the empirical and physically based models, and they use a mixture of physical and empirical equations to represent erosion processes. Hybrid models describe all of the component hydrological processes. The model parameters are assessed from field observations and also through a calibration process. Several meteorological and hydrological data are required for these calibrations (Devia et al. 2015).

10.4 ADAPTION STRATEGIES TO PROTECT COASTAL REGIONS FROM THE IMPACTS OF EROSION

Several natural and human-induced factors have contributed to increase coastal erosion. Control of these factors would be very useful to avoid or at least to reduce the negative effects of erosion processes on coastlines (Knecht and Archer 1993). In this context, many coastal erosion management techniques could be adapted and developed to protect coastal regions from the impacts of erosion:

- Controlling wind erosion:

    Various management options could be adapted to decrease wind erosion in the coastal regions. The major options include especially the increasing of soil particles cohesion using organic matter sprayed on top of the soil, the increasing of the surface of the soil's roughness by the creation of ridges of appropriate size (< 40 cm high), the colonization of coastal soil by vegetation whose roots bind sediment, making it more resistant to wind erosion, and the creation of wind-breaks: by arranging the planting of trees around a coastal area (Lebbe et al. 2008)

- Controlling wave erosion:

    For example, breakwaters (Figure 10.3) are essential to protect coastlines from wave erosion. These protective structures (usually constructed in hard materials such as concrete or rocks) contribute positively to the absorption of wave energy before the waves reach the shore, leading to significant protection of coastlines.

- Controlling sea level rise:

    This control requires serious adaptation measures such as the use of barriers such as large dams, gates, locks, or a combination of them). These barriers are highly effective tools for protecting harbors and lagoons in several countries. In addition, the coastal armoring and beach stabilization structures are widely used methods to protect the coasts from sea level rise around the world.

- Controlling human-induced factors:

    This refers to the control of land reclamation projects, river water regulation projects, land clearing activities, and mining activities. Indeed,

Climate Change and Coastal Erosion

FIGURE 10.3 Breakwater in a coastal area.

good management of these activities is critical for avoiding, or at lease reducing, adverse impacts on coastal erosion (Baric et al. 2008).

10.5 CONCLUSIONS

Coastal erosion is a particularly challenging environmental problem in all coastal regions and has historically been investigated in several natural hazard evaluation studies. This chapter has outlined the natural factors and the human-induced factors that have significantly contributed to increasing the occurrence of this environmental problem. Based on these factors, effective adaption strategies to protect coastal regions from the impacts of erosion have been proposed and discussed. In addition, the modeling approach is a useful tool for assessing climate change impacts on coastal erosion. However, the accuracy of this approach depends highly on the quality of available data (e.g. data about natural contributory factors to coastal erosion, climate data, etc.). Therefore, development of suitable methods for obtaining these data is required to evaluate accurately the effects of climate change on coastal erosion.

11 Climate Change and Water Resources Quality in Coastal Regions

11.1 INTRODUCTION

Over the past 100 years, the global air temperature has risen by 0.7 °C (IPCC 2013). Also, over this period, it was observed that the average rate of warming is about 0.13 ± 0.03 °C per decade (Rosenzweig et al. 2007). As a consequence, global warming is currently an indisputable fact (Trenberth et al. 2007). On both global and local scales, significant changes in climate variables have been identified, which include long-term changes in air temperature, rainfall, wind patterns, radiation, and increase in extreme climate events (e.g. droughts and floods) (Trenberth et al. 2007). The effects of these long-term changes on water resources have received great attention worldwide (Joehnk et al. 2008). For this reason, the potential effects of climate change on water resources have been adequately reported. Great attention was paid mainly to water resources quantity. Only a few studies have dealt with the effects of climate change on water resources quality. Research on these effects started only recently (e.g. Rosenzweig et al. 2007; IPCC 2013). These studies point out that the trends of climate change in the 20th century would have adverse effects on the quality of water resources. However, these studies did not provide details on how climate change would impact the quality of water resources (Delpla et al. 2009). Climate change can affect the quality of water resources through several biochemical processes (Dalla et al. 2007). Also, it is important to note that these effects highly depend on the type of water bodies (Whitehead et al. 2009a). A good identification of hydrodynamical and biochemical processes that occur in different types of water bodies is necessary for evaluating the relationship between climate change and water resources quality (Delpla et al. 2009). In this context, the present chapter provides a review of the observed and expected impacts of climate change on water quality in various water bodies of coastal regions and offers possible strategies for reducing these impacts.

11.2 IMPACTS OF EXTREME CLIMATE EVENTS ON WATER RESOURCES

Over recent years, several extreme climate events have been observed. These major events include storms, floods, and droughts, etc. (Tate et al. 2004).

These events have a great potential to impact the quantity and quality of costal water resources (Forbes et al. 2011). For example, studies on droughts and water resources quality (e.g. Evans et al. 2005; Ducharne et al. 2007) have revealed that extreme climate events (i.e. droughts) increased pollutant concentrations, improved nitrogen mineralization, and delayed recovery from acidification. During a drought event, lower flows have the potential to decrease the dilution process of several contaminants (Elsdon et al. 2009). The drought events that occurred in the Meuse River (Western Europe) in 2003 (compared to the year 2000), increased the concentrations of chlorophyll by about 167% due to an important reduction in the dilution impact, and the concentrations of nutrients and major ions also increased. Also, as reported in Gómez et al. (2012), sediment desiccation (as a consequence of dry conditions) has a potential to increase net nitrogen mineralization, net nitrification, and stream nitrogen availability. Furthermore, the decrease in shallow groundwater levels (as a consequence of droughts) has a potential to increase the oxidation of sulfur compounds previously stored in lakes, and re-wetting leads to the subsequent efflux of sulfate (SO_4^{2-}), which delays the recovery of surface waters from acidification (Eimers et al. 2004). Based on the outputs of some computer models such as MAGIC (Model of Acidification of Groundwater and Catchment), it was observed that under several climate change scenarios, the acid neutralizing capacity (ANC) would exhibit a critical decrease by 2080, compared to base climate scenarios (see details in Aherne et al. 2006).

On the other hand, the studies on flood and water resources quality have revealed the following points:

- Flood events have greater effects on water resources quality than drought events (Hrdinka et al. 2012);
- Together with drought events, flood events significantly change water resources quality;
- During a flood event, nitrates concentrations could be increased by critical percentages (> 100%). This high increase is mainly explained by the high concentrations of suspended solids originating from vast alluvial washouts;
- Flood events can lead to a redistribution of contaminants in the soils (Hilscherova et al. 2007);
- Flood events can lead to significant soil erosion. This in turn introduces several nutrients, pathogens, and toxins into the water environment (Cheng et al. 2007; Polyakov and Fares 2007; Ficklin et al. 2009; Wallace et al. 2009);

The international studies on the impacts of extreme climate events on water resources quality were mainly performed to reveal the impacts of droughts and floods on the concentration of contaminants. Because droughts and floods alter water and sediment conditions, which will further affect the migration and transformation of contaminants, further works must be performed to analyze the impacts of droughts and floods events on the behavior and eco-environmental effect of contaminants.

11.3 IMPACTS OF CLIMATE CHANGE ON WATER BODIES

11.3.1 IMPACTS ON LAKES AND RESERVOIRS

As revealed by several studies (e.g. Kaste et al. 2006; Wilby et al. 2006; Hammond and Pryce 2007), the impacts of climate change on lakes and reservoirs involve mainly nutrient release, growth of aquatic plants, eutrophication, and salinization. In these water bodies, the increase in water temperature, as a direct consequence of climate change, expands the thermal stratification period, deepens the thermocline and modifies the water system hydrodynamics (Jackson et al. 2007; Brooks et al. 2011). For example, as reported in Rosenzweig et al. (2007), from 1960 to 2006, the stratified period in several lakes in Northern America has been increased by 20 days. Furthermore, it was observed that the exchange of bottom water and surface water has been hindered due to thermal stratification (Delpla et al. 2009). This condition has led to a significant reduction in the concentration of dissolved oxygen and excessive carbon dioxide in the bottom water (Kock Rasmussen et al. 2009), which leads to the formation of a reductive environment (Jiang et al. 2008). Furthermore, several studies have revealed the release of nutrients and contaminants from the sediments due to bottom water hypoxia (e.g. Carvalho and Kirika 2003; Beutel 2006; Komatsu et al. 2007; Wang et al. 2008; Wilhelm and Adrian 2008; Han et al. 2009; Mihaljevi; Gantzer et al. 2009).

As reported in many studies (e.g. Mihaljević and Stević 2011; Søndergaard et al. 2011), the maintenance of ecological balance and water quality could be ensured by aquatic plants. Due to their positive contribution, the relationship between climate change and aquatic plants growth has been evaluated by several researchers (e.g. Joehnk et al. 2008; Xu et al. 2010). A lot of these studies have demonstrated that an earlier annual warming in temperate areas leads to an earlier growth of algae (Wiedner et al. 2007). Also, these studies have revealed that climate change facilitates the growth of phytoplankton in water bodies.

On the other hand, based on simulation works (a one-dimensional lake ecosystem model), Trolle et al. (2011) showed that under a climate change scenario (IPCC A2 scenario, year: 2100) the cyanophytes would be more abundant in several lakes of New Zealand by 2100. Furthermore, as reported in many studies (e.g. Cleuvers and Ratte 2002; Lloret et al. 2008), an increase in the growth of phytoplankton, as a consequence of climate change, decreases the light intensity in bottom water and leads to a significant reduction in the amounts of submerged macrophytes. Also, some studies (e.g. Ren et al. 2006; O'Farrell et al. 2011) have demonstrated that submerged macrophytes play a key role in the interception and detaining of the nitrogen and phosphorus released from sediments. Therefore, a decrease in submerged macrophytes enhances the growth of phytoplankton. But, a high density of submerged macrophytes has an important impact on the facilitation of a reductive environment. For example, Boros et al. (2011) analyzed the redox potential in a simulated shallow lake ecosystem and showed that the average redox potential was 133 ± 34 mV with a high density of submerged macrophytes and 218 ± 34 mV with a high density of phytoplankton and a low density of submerged macrophytes.

Also, aquatic ecosystems and drinking water security are affected by water salinization and mineralization. Climate change (i.e. temperature increase and rainfall decrease) is considered as a main contributor to a significant increase in lake salinization and mineralization. For example, Liu et al. (2004) have shown that an increase in temperature and a decrease in rainfall have been the main causes for the mineralization of many lakes in Northwest China since 1960. Also, Bonte and Zwolsman (2010) have shown that a temperature increase of 2 °C combined with a change in the atmosphere circulation pattern by 2050 would increase the chloride content of many lakes in the Netherlands, thereby increasing their salinization levels.

11.3.2 Impacts on Alpine Lakes

Changu Lake in India and Crater Lake in the USA are examples of these water bodies. Contrary to the lakes at lower altitudes, the alpine lakes are much less affected by discharged contaminants and anthropogenic activities (Daly and Wania 2005; Macdonald et al. 2005). However, they are fairly sensitive to global climate change (Hari et al. 2006). Several studies (e.g. Parker et al. 2008, 2008; Hruska et al. 2009) have shown that a significant increase in the concentrations of dissolved organic carbon (DOC) is a major aspect of climate change effects in alpine lakes. These studies revealed that this significant increase in the DOC is a direct consequence of important changes in the chemistry of atmospheric deposition and rises in temperature and rainfall. Furthermore, many other studies (e.g. Worrall et al. 2007; Eimers et al. 2008; Dawson et al. 2009; Hruska et al. 2009; Oulehle and Hruska 2009; Sarkkola et al. 2009) showed that changes in biodiversity are also another aspect of climate change effects in alpine lakes.

As revealed by Clarke et al. (2005), several high-altitude lakes in Europe exhibited an increasing trend in plankton diatom species over the past 150 years as a consequence of climate change. Furthermore, Fischer et al. (2011) revealed that the average Daphnia density (for the period 1991–2005) in Rocky Mountain alpine lakes had a positive correlation with increases in air temperature and DOC concentration. Also, several studies (e.g. Parker et al. 2008; Čiamporová-Zaťovičová et al. 2010) revealed that the increased air temperature, as a consequence of climate change, contributed to an important increase in the number of thermophilic aquatic insects in three alpine lakes located at different altitudes in the Tatra Mountains (Slovakia). These studies showed also that the sensitivity of aquatic organisms to climate change highly depends on the type of species.

In the alpine lakes, the impacts of climate change on water quality were also demonstrated by significant changes in the concentration of heavy metal contaminants and electrical conductivity (Kirk et al. 2011). An increase in air temperature as a consequence of climate change may affect rock weathering and glacier melting (Jing and Chen 2011), which would, in turn, affect water quality (Thies et al. 2007). For example, Tchounwou et al. (2012) investigated the lead (Pb) pollution of subarctic lakes using a Pb isotope tracer and revealed that an increase in air temperature has stimulated the transport of heavy metals from the soil to the arctic surface waters.

11.3.3 IMPACTS ON RIVER RESOURCES

These impacts include mainly nutrient loads and sediment transport, which have been investigated by several researchers (e.g. Cheng et al. 2007; Tong et al. 2007; Kaushal et al. 2008; Han et al. 2009; Tu 2009; Hamilton 2010; Lee et al. 2010; Desortová and Punčochář 2011; Wilson and Weng 2011). Zhang et al. (2012) simulated the effects of climate change on the streamflow and non-point source contaminant loads in the Shitoukou reservoir catchment (China): they demonstrated that under the A2 scenario the annual NH_4^+ load into the investigated reservoir would show a critical downward trend (average decreasing rate = 4 t year^{-1}). Also, Nõges et al. (2011) demonstrated that in the Ticino River (Europe) an increase in winter rainfall was leading to a significant increase in nutrient loadings and was causing important eutrophication. Furthermore, using the Soil and Water Integrated Model (SWIM), Martínková et al. (2011) simulated changes in the nitrate load in the Jizera catchment (Czech Republic) under the A1B scenario and demonstrated that the nitrate loads were positively correlated with the water discharge and rainfall amount. Also, based on this model (SWIM) and under the A2 and B2 scenarios, Arheimer et al. (2005) demonstrated that in the Leonard River (Sweden) a rise in the temperature would result in a 50% increase in the total phosphorus, a 20% increase in the total nitrogen, and an 80% increase in cyanobacteria.

Based on Global Climate Model (GCM) and Integrated Catchment-Sediment model (INCA-SED), Whitehead et al. (2009a) demonstrated that climate change (especially B1, B2, A2, and A1F1 scenarios) would contribute to an increase in the sediment delivery in River Lambourn (UK) during spring and autumn by 2050. This in turn would negatively contribute to an increase in storms and river flows. In addition, Verhaar et al. (2011) estimated the sediment transport from 2010 to 2099 in the Saint-Lawrence River (Canada) under B1, B2, and A2 scenarios and revealed that the sediment transport volume would show an important increase (>100%) in the investigated river under the simulated climate change scenarios. On the other hand, Prathumratana et al. (2008) evaluated the relationships between climate change and major ion content in the Mekong River (China): he demonstrated that the ion content had a negative correlation with rainfall increase from 1985 to 2004. Also, Wu et al. (2014) evaluated the relationship between the natural river runoff and the major ion content of the Yellow River (China) from 1960 to 2000 and revealed that the increased trend in the concentrations of total ions, Ca^{2+} and Mg^{2+}, in the water of this river was caused by two major factors:

- The first factor is that the increasing temperature caused by climate change enhances rock weathering, which leads to higher major ion concentration in the river water;
- The second factor is that the decreasing trends of the river runoff weaken its dilution impact for the major ions, which results in increases in their concentrations. The correlations are summarized in Table 11.1.

11.3.4 Impacts on Coastal Lagoons and Estuaries

These impacts include especially water temperature increase, nutrient loads, and eutrophication, which have been investigated by several researchers (e.g. Knowles and Cayan 2002; Yunev et al. 2007; Davies et al. 2009; Drewry et al. 2009; Whitehead et al. 2009a; Kirilova et al. 2011). Preston (2004) indicated that the water temperature of the Chesapeake Bay estuary (USA) had risen by 1.1 °C since 1950. Also, Fulweiler and Nixon (2009) showed that the average water temperature in this estuary had risen by 1.7 °C from 1988 to 2008. Li et al. (2011) investigated the nutrient input to the Yamen estuary in South China using the Soil and Water Assessment Tools model (SWAT) and revealed that a temperature rise of 3 °C would contribute to an increase in the load of sediment, inorganic nitrogen, and inorganic phosphorus of 13, 40, and 5%, respectively. Miller and Harding (2007) studied the impacts of climate change on eutrophication and showed that winters with more frequent warm or wet weather patterns are followed by a higher phytoplankton biomass in the subsequent spring, i.e., this higher biomass covers a larger area and extends into the estuary. Higher rates of primary production have also been observed in the Hudson River estuary (USA) during dry summers (Howarth et al. 2007).

TABLE 11.1
Correlations of River Runoff and Major Ion Content of the Yellow River (China) (Adapted from Wu and Xia 2014)

Measurement Method	Runoff August	Runoff September	Runoff October
Ca^{2+} content (August)	−0.338*	−	−
Ca^{2+} content (September)	−	−0.133	−
Ca^{2+} content (October)	−	−	−0.007
Mg^{2+} content (August)	−0.250	−	−
Mg^{2+} content (September)	−	−5.44**	−
Mg^{2+} content (October)	−	−	−1.25
Total ion content (August)	−0.419*	−	−
Total ion content (September)	−	−4.12*	−
Total ion content (October)	−	−	−0.074

Lloret et al. (2008) attempted to identify the role of climate change in the eutrophication of the Mar Menor coastal lagoons in Spain using a photosynthesis model and conducting experiments. They showed that the rise in the sea level as a consequence of climate change would contribute to increases in light attenuation and cause an important reduction in the amount of light reaching the water bottom, which would be associated with a loss of macroalgae and a rapid growth of phytoplankton radiation (Xia et al. 2004, 2006, 2009; Ma et al. 2010b; Fischer et al. 2011; Palmer et al. 2011; Özkan et al. 2012; Hu et al. 2012; Veríssimo et al. 2013). Solar UV radiation can penetrate to ecologically significant depths in aquatic systems, and some evidence has shown that UV radiation influences aquatic species (mainly those in estuaries), by inducing phototoxicity of some pollutants (Häder et al. 2007; Schiedek et al. 2007; Komatsu et al. 2007; Marshall and Randhir 2008; Kirilova et al. 2011; Boros et al. 2011; Gómez et al. 2012). Laboratory investigations performed by Lyons et al. (2002) have demonstrated that the toxicity of both pyrene and benzo(a)pyrene increased when the exposures were conducted in the presence of UV light.

11.4 POTENTIAL STRATEGIES FOR REDUCING THE IMPACTS OF CLIMATE CHANGE ON WATER QUALITY

Several strategies could be adapted to reduce the impacts of climate change on water quality. The major adaptation strategies include:

- According to several recent studies (e.g. Ma et al. 2010b; Özkan et al. 2012; Hu et al. 2012; Veríssimo et al. 2013), climate change can influence the sources, distribution, migration, and transformation of contaminants in various types of water bodies. Therefore, continuous control of water quality at the source and during water distribution could be helpful to identify any potential contamination;
- Carbon dioxide capture and storage, planting of bio-energy crops, proper solid waste disposition, reforestation, cropland management, both for water and reduced tillage, should be undertaken (van Vliet et al. 2013);
- A sustainable working policy on water pollution should not be only designed and enacted but also rigorously followed, more importantly in the developing and underdeveloped countries where the menace of water quality deterioration has not been effectively managed;
- Development of appropriate strategies for an effective management of contaminated water. The main objective of these strategies should be avoiding the addition of critical contaminants into the water resources;
- Some methods for removing contaminants from waters such as ozonation and advanced oxidation process (AOPs), coagulation-flocculation, membrane bioreactor, and attached growth treatment processes could be very helpful to enhance the quality of surface waters and groundwaters. They will also prevent man from further endangering his environment in the face of the menace of climate change.

11.5 CONCLUSIONS

This chapter is part of an ongoing process to enhance the understanding of the expected impacts of climate change on water quality. Climate change has several impacts on the water quality of different types of water bodies such as lakes, reservoirs, alpine lakes, rivers, coastal lagoons, and estuaries. The chapter provided the possible potential strategies for reducing the impacts of climate change on water quality. These strategies are very important to maintain the suitability of different types of water bodies for various purposes (e.g. irrigation, industry, etc.). Therefore, further work should focus on the methods used to quantify the impacts of climate change on water quality. Furthermore, research on the impacts of climate change should be extended using mechanism-based studies and experimental approaches.

12 Soil and Water Management Strategies for Climate Change Adaptation in Coastal Regions

12.1 INTRODUCTION

According to the Intergovernmental Panel on Climate Change (e.g. IPCC 2007, 2013), climate change, especially in the coastal regions, will have several pronounced effects on soil and water resources, and therefore on food security, given that soil and water have key roles in crop growth (Gudadhe et al. 2015). Furthermore, soil and water are highly sensitive to climate change due to their high vulnerability to small changes in climate variables, evapotranspiration, and sea level rise (Hossain and Lin 2001). Climatic change is expected to affect soil and water resources through several forms of degradation, including water quality deterioration, aquifer pollution, soil erosion, soil salinization, and decreased soil fertility (Cu 1993). Minimizing these degradations is the biggest and most urgent challenge facing coastal communities (IPCC 2013). Due to these degradation forms, as recommended by the Intergovernmental Panel on Climate Change (IPCC 2013), it is time to take soil and water resources seriously by identifying and evaluating suitable soil and water management adaptation strategies to cope with climate change. Therefore, in this chapter, we provide soil and water management strategies with the potential to cope with climate change and hence to reduce further degradation of soil and water resources. Furthermore, recommendations on specific soil and water use planning strategies to address climate change are provided that can be incorporated into national and international development plans. Methodologies are presented to implement these recommendations for adaptation to global climate change.

12.2 SOIL AND WATER MANAGEMENT STRATEGIES

In this section we provide all the soil and water management strategies that have the potential to cope with climate change in the coastal regions. These strategies were identified through a deep review of several international studies (more than 100 papers) that have reported impacts of climate change on coastal soil and water management (e.g. Hodny 1992; DOE 1993; Ali 1999; Hossain and Lin 2001; Cleuvers and Ratte 2002; Glassley et al. 2003; Ren et al. 2006; Lloret et al. 2008; Ng et al. 2010; O'Farrell et al. 2011; Boros et al. 2011; Trolle et al. 2011; Faour et al. 2013; McNamee et al. 2014; Rahman et al. 2015). In Table 12.1, we summarized an overview of these strategies.

12.2.1 Conservation of Key Functions of Soil and Water under Climate Change

Usually, climate change can lead to a significant disruption of the key functions of soil and water such as plant growth, water supply for agriculture, food production, carbon storage, etc. Several management actions should be implemented to conserve these functions under climate change condition (Swanston Christopher et al. 2016). Several management actions have been developed during recent years, to reduce potential climate change impacts, and these are discussed below.

12.2.1.1 Enhance Soil Health

Healthy soils are essential to assure perfect productivity of agricultural fields and for conserving the key functions of ecosystems such as biological activities, enhancing water quality, providing micro and macro nutrients for crops, and ensuring carbon sequestration (Freidman et al. 2001). Soil physical, chemical, and biological properties have significant effects on these key functions. Many of these soil properties are dynamic and have a high potential for alteration. Contrarily, a few soil properties are inherent and more resistant to alteration. Through direct and indirect ways, climate change has various negative effects on soil health, which in turn lead to important challenges for agricultural productivity and sustainability. Several soil management actions can help

TABLE 12.1
Overview of Soil and Water Management Strategies Under Climate Change

Soil or Water Strategy	Actions Examples
Enhance soil health	Continuous increase of soil organic matter Crop rotation diversification Land leveling
Conserve water quality	Decrease nutrients in lakes and streams Manage nutrients in agricultural fields Install controlled drainage systems
Manage aquifer recharge	Infiltration mechanisms Direct injection mechanisms Filtration mechanisms
Reduce soil erosion	Cut-off drains Retention ditches Water-retaining pits

us to improve soil health and reduce the negative effects of climate change. For example, the negative effects of dry and wet rainfall extremes can be reduced by increasing soil organic matter which can enhance water infiltration and decrease nutrient losses during extreme rainfall events (FAO 2007; Anwar et al. 2013). Also, green infrastructures such as pipes and drains can reduce soil erosion and also may enhance inherent soil properties. The major soil management actions that can help to reduce the negative effects of climate change on soil health include:

- Soil management action 1:

  Yearly increase of vegetation cover (by adding residue of plants) can significantly help to decrease soil exposure to water and wind erosion (Derner et al. 2015);

- Soil management action 2:

  Continuous increase of soil organic matter (by adding organic amendments such as animal manure, compost, mulch, biochar, etc.) can significantly help to enhance several properties of soil (e.g. soil water-holding capacity, soil structure, water infiltration, etc.). It helps also to decrease soil exposure to water and wind erosion (Shea 2014);

- Soil management action 3:

  Crop rotation diversification (i.e. the practice of different plants and crops in the same field) can significantly help to enhance below-ground conditions for soil life and avoid disease, weeds, and insect pests;

- Soil management action 4:

  Land leveling, subsurface drains, and perennial crop land use systems are so useful to conserve the physical, chemical, and biological properties. These practices help also to reduce crop damage from water ponding after a heavy rainfall event, and to control runoff without causing soil erosion. This leads to perfect crop growth and sustainable agricultural productivity (Ritzema 1994);

- Soil management action 5:

  Good management of soil before crop planting and good selection of planting dates could significantly help to avoid field work under extreme weather conditions such as wet conditions (Wolfe et al. 2011);

- Soil management action 6:

  Preventing soil compaction by equipment through avoiding work under wet conditions, reducing tillage, and use of the right implements contributes positively to conserving the physical properties of soil (DeJong-Hughes et al. 2001). Soil compaction can be reduced through better soil management such as the promotion of organic matter in the soil;

- Soil management action 7:

  Good planning of grazing animals greatly enhances the biological properties of soil (Bardgett et al. 1997). Grazing animals also have a good potential for enhancing soil structure, and increasing the capacity of soil organic carbon storage that could increase nutrient retention, water storage, pollutant attenuation, soil fertility, and crop productivity (Hassink 1994);

- Soil management action 8:

  Installation of windbreaks (Figure 12.1) in coastal agricultural areas, especially in those near the coastline could be a perfect adaptation strategy to decrease wind erosion. Furthermore, colonization of coastal soil by vegetation whose roots bind sediment could be so useful for more resistance to wind erosion (Lebbe et al. 2008).

  Installation of windbreaks in agricultural fields has a variety of benefits such as protection of wind sensitive crops, improving water use efficiency by lowering soil evaporation rates across protected areas, increasing water holding capacity, and increasing crop production, etc. (see Figure 12.2).

- Soil management action 9:

  Rotational grazing system (i.e. the practice of moving livestock between pastures on a regular basis) instead of traditional grazing system allows

FIGURE 12.1 Windbreaks in North Dakota, USA (adapted from Wright and Stuhr 2002.)

Soil and Water Management Strategies

FIGURE 12.2 Main benefits of windbreaks in an agricultural field.

crops to continually produce large volumes of high quality leaf material (Gerrish 2004). This can lead to optimum agricultural productivity.

- Soil management action 10:

  Managing rates and timing of harvesting hay, biomass, or other similar herbaceous perennial crops could be so useful to enhance soil health (Hodgson and Illius 1996).

12.2.1.2 Conserve Water Quality

Water resources such as coastal aquifers, lakes, and streams are vital to all agricultural activities in coastal regions. These agricultural activities under climate change condition can greatly affect the quality of water resources. Climate change has wide effects on agricultural production in coastal areas, depending on the agricultural area location, the type of agricultural system, and the type of change. For example, extreme rainfall events (i.e. increases in rainfall amount) during some times of the year have a high potential to increase runoff of fertilizers or effluents. This leads to increases of nitrogen or other nutrients in downstream water bodies (Ames and Dufour 2014). Consequently, it is so urgent to find specific practices in the agricultural fields that can conserve water quality and respond to extreme climate events. For all coastal agricultural fields, it is recommended to follow the following management actions:

- Management action 1: decrease nutrients in lakes and streams

  For example, the deep-rooted plants along streams and lakes have a great absorption capacity to decrease nutrients (Blanc and Reilly 2015);

- Management action 2: manage nutrients in agricultural fields

    Good nutrient management plans are so required to ensure perfect use of all sources of nutrients, especially under climate change condition. For example, don't apply pesticides unless there is an identified need. This leads to optimum protection of water resources (Howden et al. 2007);

- Management action 3: install controlled drainage systems

    Installation of these controlled drainage systems instead of traditional pattern tiling is so useful to avoid pollution of aquifers, especially pollution of shallow groundwaters (Haj-Amor et al. 2018);

- Management action 4: strict respect of manure management plans

    Follow manure management plans, including setbacks from water resources when applying manure to fields (Blanc and Reilly 2015);

- Management action 5: limit seawater intrusion

    Several practices can help to limit seawater intrusion. The major practices include continuous measurement of coastal groundwater level, decreasing abstraction from shallow groundwaters, installing barriers to fluvial saltwater intrusion, and increasing sustainable aquifer recharge (Krupavathi and Movva 2016);

- Management action 6: increase irrigation capacity

    Under dry conditions and where soils have low infiltration rates, increasing irrigation capacity, especially for high-value crops, is so essential to carry the salts from the soil out to drainage networks (Walthall et al. 2012);

- Management action 7: increase irrigation efficiency

    Increasing irrigation efficiency (especially water application efficiency within the irrigated fields) with the latest irrigation technology such as micro or drip irrigation systems is so useful to ensure sustainability of irrigation water resource (Haj-Amor et al. 2018);

- Management action 8: apply less irrigation in non-drought years

    Applying less irrigation in non-drought years (so saving water for further use during drought years) is also so useful to ensure sustainability of irrigation water resource (Derner et al. 2015);

- Management action 9: intensify use of technologies

    Intensifying use of technologies to "harvest" water, conserve soil moisture (e.g., crop residue retention), and using and transporting water more effectively where precipitation is reduced (Howden et al. 2007);

- Management action 10: use new technology

    Use new technology for subsurface irrigation, and irrigate with gray or reclaimed water to reduce water use (Derner et al. 2015);

- Management action 11: minimize impacts of agricultural waste

  Minimize various impacts of agricultural waste on surface and groundwater resources. For example, it is essential to respect the appropriate distance from any watercourse (e.g. 20 m) when applying fertilizers and organic wastes (Howden et al. 2007).

12.2.2 Managed Aquifer Recharge

Managed aquifer recharge refers to the intentional storage of water into an aquifer for subsequent recovery (i.e. for later uses). Managed aquifer recharge is mainly practiced to store and treat water in a suitable aquifer from various sources such as river water, rainwater, storm water, desalinated seawater, and water from other aquifers (Dillon and Pavelic 1996). The stored water (i.e. after aquifer recharge) can be used for several purposes such as drinking water, industry, and irrigation (Hammer and Elser 1980). However, before these uses, appropriate pre-treatment prior to recharge and post-treatment after recovery are required (see Figure 12.3).

Because of continued growth of coastal communities, increasing pressures on water resources, and climate change, safe and reliable storage, and treatment of water in the subsurface have received great attention (Murray 2008). The wide range of applications of the various aquifer recharge techniques are summarized in Figure 12.3. It is important to note that aquifer recharge in coastal regions has several benefits such as enhancing coastal water quality by reducing urban discharges, reducing floods and flood damage, and facilitating urban landscape improvements that increase land value (Ahmed et al. 2001). There are several mechanisms that can ensure aquifer recharge. The major mechanisms include infiltration basins and galleries for rainwater, injection wells, and filtration mechanisms (Figure 12.4). With each recharge mechanism, many factors can influence the performance of aquifer recharge, and there are many associated environmental risks. In the following sections, we discuss in detail the main mechanisms of aquifer recharge, the factors influencing the performance of aquifer recharge, and then we summarize the major environmental risks for aquifer recharge projects.

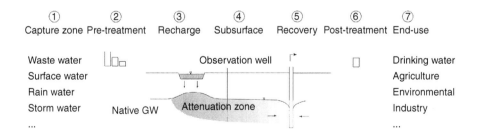

FIGURE 12.3 Main components of managed aquifer recharge from the source to the end-use.

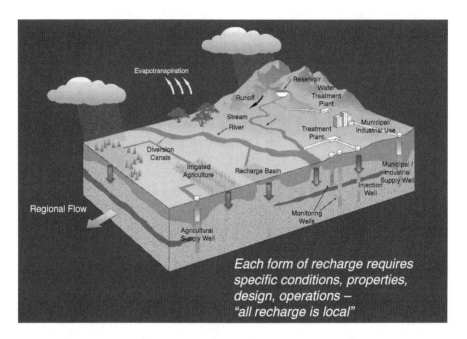

FIGURE 12.4 Main mechanisms of aquifer recharge: recharge basins, injection wells, etc. (adapted from Fisher et al. 2017.)

12.2.2.1 Mechanisms of Aquifer Recharge

- Infiltration mechanisms:

  These mechanisms improve the infiltration of rainwater and pumped water to the aquifer through infiltration basins such as ponds and tanks (Al-Zubari 2003). In the infiltration basins, infiltration rates should be continuously measured to detect any decrease in water infiltration in relation with clogging (Armandine Les Landes et al. 2014). These mechanisms enhance the quality of the recharge water based on natural attenuation in the unsaturated zone of the aquifer (Azaroual et al. 2008). Often, the infiltration mechanisms are applied to recharge shallow coastal aquifers or, to install hydraulic barriers (Azaroual et al. 2009). The infiltration mechanisms have many benefits; however, the main benefits are that they are easy to install in the field and they have low cost compared to other mechanisms (Bekele et al. 2011). In addition to the infiltration basins such as ponds and tanks (that include installations such as dams and small reservoirs), the infiltration mechanisms also include reservoir pavements, recharge pits, drainage trenches, vegetated ditches, mounds systems, sand filters, and septic drain fields (Chafidz et al. 2014). All these methods contribute positively to effective purification of recharge water due to the sufficient time for water percolation through the soil before reaching the aquifer (Dabbagh 2001). Water percolation results in avoiding pathogenic agents and several harmful inorganic and organic substances (De Montety et al. 2008). Usually, the infiltration basins are installed in areas that show important water shortages. The properties of these infiltration basins (e.g. basin size, infiltration rate, etc.) depend highly on the methods used, such as infiltration ponds, percolation tanks, etc. (Dillon 2005). Furthermore, these properties should be strictly adapted to the local objectives (Dillon et al. 2006). For example, if the objective is quantitative, the selected infiltration rate should be high (e.g. 20 m day^{-1}). However, if the objective also involves geo-purification of the recharge water, the selected infiltration rate should be low (e.g. 0.2 m day^{-1}). The geo-purification capacity of the soil in many cases is enhanced when some crops (especially crops with deep root-zone) are planted on this soil. Indeed, the presence of these crops in the soil protects the surface of the infiltration basins from erosion and clogging and provides bacteria that biodegrade many inorganic and organic pollutants. Furthermore, these crops can lead to effective purification of the recharge water by improving phytoremediation (Feitelson and Rosenthal 2012). The infiltration rate also depends on water temperature. Contrarily to the water with high temperatures, water with low temperatures has a great ability to infiltrate slowly due to an increase in viscosity (Foster et al. 2013). Consequently, the volume of water that infiltrates below a basin can reduce in winter. Water that has not been correctly treated is usually characterized by important organic matter, which fosters the development of bacteria. This can lead to significant reductions in soil porosity, especially by the formation of biofilms (Giuntoli et al. 2013). To avoid the occurrence of the clogging process in the infiltration basins, many procedures should be performed. The first procedure is the pre-treatment of the recharge water. In this procedure, sand filters are installed upstream from the infiltration basin, the chemical characteristics of the recharge water is modified by some inorganic compounds before going to the infiltration basin (Gleeson et al. 2012). In the second procedure, we can operate infiltration basins alternately (following the "wetting–drying" cycle). This procedure leads to effective decompacting of the soil in the infiltration basin (Haeber and Waller 1987).

- Direct injection mechanisms:

The injection wells are common mechanisms applied in many regions of the world (Hunter et al. 1998). The wells are mainly installed to achieve two objectives: firstly, to recharge confined or semi-confined aquifers, especially in coastal regions and secondly to install hydraulic barriers (IPCC 2014). In these mechanisms, the quality of the injected water should be highly investigated to avoid any pollution. Furthermore, they should be preferred when the area is small (Johnson et al. 1999). The direct injection mechanism refers to the injection of water into an aquifer followed by its recovery by pumping from the same well at a later date (Khan et al. 2008). This technique allows storage of excess water during the rainy periods and the consumption of this stored water during periods of water shortage. The direct injection mechanism is usually used for aquifers that are invulnerable to non-point source contamination and in which groundwater moves slowly (as is the case of confined or semi-confined aquifers). This mechanism is often applied to save water with good quality (usually potable water). A major advantage of direct injection mechanisms is that they entail alternating phases of injection and abstraction in the same well. This leads to an inversion of the water circulation in the well screen and in the surrounding aquifer, which results in a significant decrease in clogging (Dillon et al. 2006). The second advantage refers to the low cost of the direct injection mechanisms due to the use of the same well for injection (Kloppmann et al. 2012).

- Filtration mechanisms:

  These mechanisms are also referred to as "riverbank filtration mechanisms" because their main objective consists of increasing the infiltration of water from a river to its alluvial aquifer by pumping in wells located close to the riverbank (Ge et al. 2010). Therefore, several wells are installed parallel to and close to the riverbank. Pumping in these wells decreases water table levels and creates a significant difference in head between the river and the groundwater. Due to the good geo-purifying capacity of the riverbank, the filtration mechanisms lead usually to effective filtration and purification of the stored water. Because of the high concentration of suspended matter in surface water, the clogging process rapidly occurs in the riverbanks. Therefore, it is recommended that the infiltration rate should be low and the riverbanks should be continuously maintained to avoid the occurrence of the clogging process. In addition, dune filtration (i.e. infiltrating water from ponds constructed in dunes) can help to avoid the occurrence of the clogging process (Dillon 2005).

12.2.2.2 Factors Influencing Performance of Aquifer Recharge

The performance of aquifer recharge depends on several factors. The major factors include hydrogeological factors, hydrological factors, and climate conditions:

- Hydrogeological factors:

  The feasibility of aquifer recharge depends on several hydrogeological parameters such as porosity of soils, hydraulic conductivity of the aquifer, transmissivity, etc. These parameters can help to select the optimum location and suitable structure of sites of aquifer recharge projects. The main objective is to identify aquifers that store large quantities of water and not release them too quickly (Rinck-Pfeiffer et al. 2000). Due to the great importance of these parameters, detailed investigation of some hydrogeological parameters is required to install successfully the aquifer recharge projects (Roux 1995):

- Geological and hydraulic boundaries: Regional geology and hydrogeology maps of the project site could be so helpful to get these parameters;
- Inflow and outflow of waters: these parameters could be measured directly in the project site or also could be obtained from environmental/water authorities;
- Storage capacity, porosity, hydraulic conductivity, and transmissivity parameters: pumping tests and hydraulic flow models are helpful to estimate correctly these important data. The measured and modeled data must be carefully interpreted and checked because they contribute directly to the selection of the optimum location and suitable structure of sites of aquifer recharge projects;
- Natural discharge and recharge: these parameters are calculated based on the water table fluctuation method or the hydrologic budget method. Remote sensing-GIS approaches can be also so helpful to get natural discharge and recharge of aquifers;
- Water availability for recharge and water balance: source water availability may be determined from the annual rainfall data, river flow estimation, surface runoff estimation, etc.;
- Lithology, depth of the aquifer, and tectonic boundaries: these data are estimated from boreholes, aerial photographs, etc.

- Hydrological factors:

  Like those of the hydrogeological parameters, the hydrological factors are also critical for selecting suitable areas for aquifer recharge projects and also for estimating the potential amount of water available for recharge (Wintgens et al. 2012). Good selection of these areas will help reduce the implementation and operational costs of aquifer recharge projects. There are several hydrological factors that must be taken into account during the implantation of an aquifer recharge project; however, the most important factors are:

  - Area properties (especially topography, elevation, and slope);
  - Land management (agricultural practices, urban areas, barren land etc.);
  - Vegetation properties (such as type of crops, vegetation cover thickness, etc.);
  - Flow availability and rate in streams (perennial, ephemeral, large/small rivers);

○ Conveyance system for bringing the water (e.g. gravity flow, pumping, etc.).

- Climate conditions:

  During the design of an aquifer recharge project, climactic conditions play a key role in the determination of the dimensions and type of structures that need to be implemented. The major climate variables that should be determined include average annual rainfall (so important for determining the size of the structure), maximum rainfall intensity, number of rain days, temporal variation in air temperature, evaporation, evapotranspiration, and shifts in seasonal patterns (alternate dry and wet season, and shallow groundwater fluctuations).

12.2.2.3 Environmental Risks of Aquifer Recharge Projects

As reported by Lapworth et al. (2012), many contaminants (e.g. trace metals, nutrients, pathogenic microorganisms, contaminants of emerging concern) can be present in recharge water due to the use of a variety of water sources (e.g. collected rainwater, treated wastewater, etc.) for aquifer recharge. The performance of the pre-treatment process of water before the aquifer recharge contributes to the determination of the concentrations of these contaminants. Effective pre-treatment of water before the recharge leads to insignificant contamination. The presence of these various contaminants has an important potential to create high risks for public health. The highest risk usually occurs when the treated wastewater is selected for aquifer recharge. This risk is assessed based on numerical simulation of reactive transport processes in the unsaturated zone of the aquifer. Usually, two major challenges must be taken into consideration if treated wastewater is being considered for aquifer recharge. The first challenge is the integration of all hydro-biogeochemical data in the reactive transport processes (i.e. during numerical modeling). The second challenge is the need for a full description of various hydrogeological factors specific to each aquifer recharge site.

- Risks of trace metals:

  Trace metals such as iron, zinc, copper, manganese, chrome, arsenic, selenium, mercury, cadmium, molybdenum, and nickel can be dissolved in recharge water from the aquifer material by great modification of natural geochemical conditions. Naturally (i.e. in natural water), the concentrations of trace metals are insignificant. The increase in the concentrations of these trace metals in the recharge water is mainly attributed to the use of treated wastewater for aquifer recharge. This increase could have various health and environmental risks. Accordingly, it is recommended that trace metal concentrations must be below acceptable levels (Haeber and Waller 1987). Some trace metals, especially iron, zinc, copper, and lead could be also presented in the recharge water which coming from water treatment plants. Furthermore, other trace metals (e.g. manganese, aluminium, chrome, arsenic, selenium, mercury, cadmium, molybdenum, and nickel) could also be present due to several causes and origins:

- From the corrosion of material in the water distribution and treatment systems;
- From service activities (such as health, automobiles, etc.);
- From industrial effluents (such as petrochemical industry discharges, etc.).

The presence of trace metals in the recharge water with high concentrations typically leads to the creation of new redox conditions in the system driven by the microbial community. Many countries, such as the USA, Australia, France, and Germany have developed guidelines for the use of treated wastewater for recharge (USEPA 2012). The main objective of these guidelines is to define adequately the maximum concentrations of trace metals in the recharge water that can help to avoid the health and environmental risks of trace metals. The following recommendations are examples of these guidelines:

- In France, the maximum concentration of arsenic in drinking water is 10 μg l^{-1};
- Nickel, which is weakly toxic but which accumulates in plants;
- Acceptable daily intake (ADI) of cadmium is 0.057 mg/day/individual;
- Mercury, this trace metal can be highly mobile, so it is recommended to greatly reduce the concentration of this element in recharge water.

- Risks of emerging contaminants

Currently, the quality of recharge water is highly threatened by various emerging contaminants and pathogens such as antibiotic resistant bacteria and viruses (Zwiener et al. 2002). Often, the treated wastewater used for aquifer recharge is highly contaminated by these emerging contaminants, especially by pharmaceutical products. Indeed, there are many origins of pharmaceutical products discharged to water bodies. The pollution of wastewater is usually attributed to the excretion of pharmaceutical products by patients following their ingestion. Consequently, hospital wastewater contains high levels of pharmaceutical products, mainly antibiotics. Furthermore, some other products such as anesthesia products and diagnostic products are also present in wastewater at relatively high levels. Many studies have identified pharmaceutical products in treated wastewater (e.g. Steger-Hartmann et al. 1996; Ternes et al. 1998). These studies revealed that the most common pharmaceutical products in wastewater are beta blockers, contrast media, and pain relief/anti-inflammatory drugs. These pharmaceutical products, especially in their active forms, can have several effects on health and the environment. Important eco-toxicological effects are usually observed.

During the treatment of wastewater, the destruction of the pharmaceutical products varies depending on both the nature of the drug under consideration and on the properties of the treatment processes used in the treatment plant (Joss et al. 2005; Yu et al. 2006). Similar to the trace metals, infiltration basins and indirect injection methods would help to decrease the mobility of pharmaceutical products. In these methods, the presence of an unsaturated zone improves trapping of the pharmaceutical contaminants. Geochemical and microbiological processes such as biodegradation and adsorption occurring in the unsaturated zone have a great potential to reduce the concentrations of pathogenic and non-pathogenic microorganisms and emerging contaminants. The adsorption process also can potentially reduce the mobility of organic contaminants. The adsorption of pharmaceutical products usually occurs on many solid phases in the unsaturated zone (e.g. oxyhydroxides, clay minerals, and humic substances). The creation of a chemical link between the functional groups present on a pharmaceutical product and the functional groups present on these solid phases is required to ensure effective adsorption of pharmaceutical products.

- Environmental risk assessment

  The environmental risks of aquifer recharge projects have been assessed by several international studies (e.g. Wintgens et al. 2012). These studies revealed three categories of environmental risks: expected theoretical risks, expected experimental risks, and real environmental risks. The first category (i.e. expected theoretical risks) refers to all of the disruptions that might affect the properties of the aquifer as a consequence of the installation of an aquifer recharge project. These disruptions include especially aquifer pollution by recharge water that shows high levels of contaminants such as pharmaceutical products and trace metals. Consequently, the assessment of expected theoretical risks needs the identification of:

  - All natural potential origins of pollution of the recharge water used. The main natural origins include especially the presence of areas rich in trace metals;
  - All chemical and microbiological properties of the recharge water;
  - All human contributors to aquifer pollution such as industrial effluents, presence of a nearby hospital, etc.

  The expected theoretical risks refer to the conveyance of a contaminant from the recharge water to humans and the environment. Under this condition, in addition to the theoretical factors, the expected experimental risks will depend on other factors such as the volumes of recharge water injected, the performance of pre-treatment processes, and especially the geo-purification capacity of the unsaturated zone of the aquifer (as in the case of infiltration basin and indirect injection methods). Finally, the real environmental risks refer to the probability that an exposed population will be contaminated. Contrary to the theoretical and experimental risks, the real environmental risks consider various factors that are specific to a population exposed to disruptions resulting from aquifer recharge projects such as age, sex, health, nutrition, and specific immune system capacity of a given individual. Finally, it is important to note that, despite the positive outcomes of several risk assessment studies, especially the development of a conceptual framework of the environmental risks caused by aquifer recharge projects, the weak identification of some contaminants increases the risks associated with the disruptions caused by these projects.

12.2.3 Measures for Reducing Soil Erosion

The coastal regions are the most vulnerable regions of the world to destructive soil erosion (Fedorova et al. 2010) because of several factors such as sea level rise, storm events, and storm surges that are occurring more frequently as a consequence of climate change (Ferreira et al. 2009). Destructive coastal erosion has various negative impacts on farmland, the coastal ecosystem, and rural infrastructure such as roads and sea walls (Mangor et al. 2017). Due to these impacts, coastal erosion has received wide attention (e.g. Vestergaard 1991; Tõnisson et al. 2008; Soomere et al. 2011). Hence, several physical soil conservation structures were designed to conserve the areas suffering from soil erosion. These structures have also the potential to retain water where needed (Gill et al. 2008). These structures have a variety of benefits such as decreasing the velocity of surface runoff, increasing soil moisture, maintaining good soil cover through mulching and canopy cover, enhancing soil structure for decreasing crusting, enhancing soil fertility, and avoiding excessive runoff safely. Several factors have contributed to the design of physical soil conservation structures including climate conditions, size of agricultural field, soil properties (especially texture, infiltration capacity, and depth), runoff properties, availability of an outlet, soil management options within the agricultural field (e.g. vegetative conservation measures), and labor availability and cost (Carluer and Marsily 2004). The common physical conservation measures are:

- Cut-off drains (Figure 12.5): they are made across a slope for intercepting the surface runoff and carrying it safely to an outlet (e.g. a canal or stream). The objective of these drains is to protect agricultural areas, compounds, and roads from uncontrolled runoff, and to divert water from gully heads;
- Retention ditches: Usually these structures are designed for water harvesting in semi-arid areas (Gallart et al. 1994). These structures lead usually to a perfect prevention of soil erosion by water in agricultural fields. This prevention consists of limiting the erosion capacity of soils and

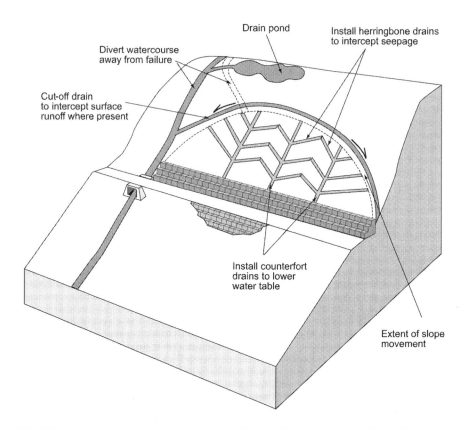

FIGURE 12.5 A schematic representation of cut-off drains (adapted from Hearn and Hunt 2011.)

decreasing the intensity of surface runoff, which is a main contributor to the detachment and transport of soil particles (Childs and Youngs 2006). Usually, these structures are installed on the margins of agricultural fields. This installation contributes positively to a significant surface runoff decrease (Dunn and Mackay 1996). These structures are used to reduce the slope length throughout an agricultural field by intercepting and channelizing the runoff waters (Levavasseur et al. 2012);

- Infiltration ditches: often, these ditches are designed to harvest water from roads or other sources of runoff. They comprise digging along the contour, upslope from a crop field, a ditch of around 1 m deep. This helps stop down-slope water movement as the water falls into the ditch. These structures are easy to build, and if constructed at a 1% slope, they divert excess water to protected drainage ways, highly decreasing soil erosion and leaching of nutrients (Govers et al. 2004);

- Water-retaining pits: these structures refer to the installation of a series of pits that are dug into the ground where runoff normally occurs. Then, the soil from the pit is used to make banks around the pits. Furrows carry excess water from one pit to the next. This technique helps to slow the flow of water within the agricultural field (by trapping the water). The size of pit depends mainly on the amount of runoff. Usually, the size is 2 m square and 1 m deep (Sun 2002). The installation of water-retaining pits within an agricultural field is simple and inexpensive. These structures are easily understood and adopted by farmers. However, some theoretical problems such as infiltration of varying head, complicated boundary in storage pit, and its mathematical modeling etc. should be investigated further to estimate correctly the technical parameters of water-retaining pits (Ma et al. 2010b);
- Gabion walls: These structures are designed to stabilize the soils behind the wall (see Figure 12.6). This technique is considered as the most economical and efficient strategy to stabilize natural ground slope. Furthermore, Gabion walls are preferred due to their flexibility, permeable nature, low cost, and their conservation of the environment (Kandaris 1999).

FIGURE 12.6 A photo of a Gabion wall (adapted from Chen and Tang 2011.)

12.3 CONCLUSIONS

Agricultural productivity greatly depends on soil and water resources. High agricultural productivity is achieved critically by conserving these key resources. In this chapter, we discussed the basic soil and water management strategies that have the potential to cope with climate change and hence may reduce further degradation of soil and water resources. Furthermore, recommendations on specific soil and water use planning strategies to address climate change were provided that can be incorporated into national and international development plans. Degradation of soil and water resources is reduced and effectively stabilized in the coastal regions by the implementation of suitable soil and water management strategies as summarized in this chapter. However, as always, much more needs to be done. Poor farmers that do not have adequate resources to implement these strategies and mitigate the threat of soil and water degradation in developing countries require greater attention.

General Conclusions

Coastal regions are very rich in soil and water resources with ecological, economic, and social significance. These regions are fragile and vulnerable to climate change as well. The coastal regions of the world are typically different from each other with respect to climate conditions, geography, and agro-ecological conditions. However, all coastal regions are suffering from climate change. Addressing the issues pertaining to natural soil and water resources and agricultural production in the coastal regions requires assessment of the existing resources. This book analyzed properties of coastal soil and water resources, climate conditions, climate change aspects, water resources deterioration, and land degradation processes. Based on the findings of this book, coastal regions are highly affected by climate change, and these impacts will continue to affect coastal areas in a variety of aspects. These aspects include rising atmospheric concentrations of carbon dioxide (CO_2), increased air temperature, changes in rainfall patterns, droughts, floods, ocean acidification, and sea level rise. Furthermore, various soil management practices such as agroforestry, cultivation of perennial crops, and reducing tillage, have a great potential to sequester carbon and reverse the carbon enrichment of the atmosphere. However, this potential of soil carbon sequestration may be restricted in practice due to landscape variability and the effects of atmospheric change and climate warming. Despite these restrictions, the benefits of increasing soil organic carbon sequestration, even on a limited basis, are very important for maintaining and enhancing agricultural productivity and supporting vital ecosystem services and should be supported through advanced research initiatives.

The effects of climate change are likely to worsen several environmental problems that coastal regions already face. Shoreline erosion, coastal storms, and flooding have strong negative effects on agricultural productivity and coastal ecosystems. In addition, contamination of water resources via the climate change effect of seawater intrusion makes the coastal regions less resilient to climate impacts. Confronting existing challenges is already a concern. Addressing the additional stress of ongoing climate changes will require identification of suitable strategies to manage water resources well, to avoid land degradation, and hence to conserve coastal ecosystems. From Chapter 10 to Chapter 12, the book summarized the key adaptation options to minimize the impact of climate change on soil and water resources in coastal regions. A lot of soil and water adaptation strategies were presented and discussed.

It is important to note that the soil and water measures reported in this book are considered as being helpful activities at the local level (i.e. within the field) to maintain and improve the productive capacity of the soil and water resources and vegetation in areas prone to degradation through prevention or decreasing of soil erosion, soil salinization, and water pollution, and through conservation of drainage water and enhancement of soil fertility. These measures are to be selected and implemented according to the respective local conditions;

i.e., the strategy is adapted at the local level. For this reason, in this book recommendations on specific soil and water use planning strategies to address climate change were provided which can be incorporated into national and international development plans. Also, methodologies were presented to implement these recommendations for adaptation to global climate change.

Finally, it important to note that further work must be done in order to estimate correctly and adequately the technical parameters of soil and water conservation measures in coastal regions. In this regard, modeling (i.e. computer programs) would be very helpful. Introducing correct and adequate soil and water conservation measures in a coastal area may further the sustainable utilization of natural resources for the benefit of local people. In addition to the technical parameters, the success of soil and water conservation measures will depend on the participation of local people with their traditional knowledge. Coupled with technical parameters, this traditional knowledge will help implement and carry out the soil and water conservation measures suitable for the natural and human conditions of the coastal area. Several economic, institutional, and political aspects have to be considered as well. Furthermore, environmental education of the involved public and capacity building of regional actors is an important cornerstone for the success of soil and water conservation measures.

References

Abbot, D.S., and I. Halevy. 2010. Dust aerosol important for snowball earth deglaciation. *J Clim* 23:4121–4132.

Abbott, M.B., J.C. Bathurst, J.A. Cunge, P.E. O'Connell, and P. Rasmussen. 1986. An introduction to the European hydrological system, 1: History and philosophy of a physically-based, distributed modelling system. *J Hydrol* 87:45–59.

Abrol, I.P., R.K. Gupta, and R.K. Malik 2005. *Conservation Agriculture: Status and Prospects.* Centre for Advancement of Sustainable Agriculture, New Delhi. p. 242.

Adelana, S., Y. Xu, and P. Vrbka. 2010. A conceptual model for the development and management of the cape flats aquifer, South Africa. *Water SA* 36:4.

Adelana, S. M., A. M. MacDonald, S. Adelana, and A. MacDonald. 2008. Groundwater research issues in Africa. In: *Applied Groundwater Studies in Africa. IAH Selected Papers on Hydrogeology*, Adelana, S. A. M. and A. M. MacDonald (Eds.) Vol. 13:1–7.

Admiraal, W., and Y.J.H. Botermans. 1989. Comparison of nitrification rates in three branches of the lower river Rhine. *Biogeochemistry* 8:135–151.

Aherne, J., T. Larssen, J.B. Cosby, and J.P. Dillon. 2006. Climate variability and forecasting surface water recovery from acidification: Modelling drought-induced sulphate release from wetlands. *Sci Total Environ* 365(1–3):186–199.

Ahmad, P., and S. Sharma. 2008. Salt stress and phyto-biochemical responses of plants. *Plant Soil Environ* 58:89–99.

Ahmed, M., W.H. Shayya, D. Hoey, and J. Al-Handaly. 2001. Brine disposal from reverse osmosis desalination plants in Oman and the United Arab Emirates. *Desalination* 133(2):135–147.

Ajayi, O.C., F. Place, F.K. Akinnifesi, and G.W. Sileshi. 2011. Agricultural success from Africa: The case of fertilizer tree systems in southern Africa (Malawi, Tanzania, Mozambique, Zambia and Zimbabwe). *Int J Agric Sustain* 9:129–136.

Al Charaabi, Y., and S. Al-Yahyai. 2013. Projection of future changes in rainfall and temperature patterns in Oman. *Journal of Earth Science and Climate Change* 4:154.

Alabi, A.A., R. Bello, A.S. Ogungbe, and H.O. Oyerinde. 2010. Determination of groundwater potential in Lagos State university, Ojo; using geoelectric methods (vertical electrical sounding and horizontal profiling). *Report Opinion* 24:68–75.

Alfaro, S.C. 2008. Influence of soil texture on the binding energies of fine mineral dust particles potentially released by wind erosion. *Geomorphology* 93:157–167.

Alfieri, L., P. Claps, P. D'Odorico, F. Laio, and T.M. Over. 2008. An analysis of the soil moisture feedback on convective and stratiform precipitation. *J Hydrometeorol* 9(2):280–291.

Ali, A. 1999. Climate change impacts and adaptation assessment in Bangladesh. *Clim Res* 12: 109–116.

Ali, R., D. McFarlane, S. Varma, W. Dawes, I. Emelyanova, and G. Hodgson. 2012. Potential climate change impacts on groundwater resources of south-western Australia. *J Hydrol* 475:456–472.

Allen, D.M., D.C. Mackie, and M. Wei. 2004. Groundwater and climate change: A sensitivity analysis for the grand forks aquifer, southern British Columbia, Canada. *Hydrogeol J* 13:270–290.

Almeida, L.P., M.V. Vousdoukas, O. Ferreira, B.A. Rodrigues, and A. Matias. 2012. Thresholds for storm impacts on an exposed sandy coastal area in southern Portugal. *Geomorphology* 143–144:3–12.

Al-Zubari, W.K. 2003. Assessing the sustainability of non-renewable brackish groundwater in feeding an RO desalination plant in Bahrain. *Desalination* 159(3):211–224.

Ames, G.K., and R. Dufour, 2014. Climate change and perennial fruit and nut production: Investing in resilience in uncertain times. In: *ATTRA Sustainable Agriculture*, Amy, S. (Ed.) National Center for Appropriate Technology ATTRA, Butte, MT.

Amin, S.M.N., and R.G.D. Davidson-Arnott. 1995. Toe erosion of glacial till bluffs: Lake Erie south shore. *Can. J. Of Earth Sci.* 32:829–837.

Anderson, L. 2003. Evaluation of shoreline erosion extent and processes on Wisconsin's Lake Superior shoreline. M.S. thesis, University of WisconsinMadison, Madison, Wisconsin.

Anwar, M., D. Liu, I. Macadam, and G. Kelly. 2013. Adapting agriculture to climate change: A review. *Theor Appl Climtol* 113(1–2):225–245.

Apitz, S.E., J.W. Davis, K. Finkelstein et al. 2005. Assessing and managing contaminated sediments: Part I, developing an effective investigate on and risk evaluation strategy. *Integr Environ Asses* 1:2–8.

Arheimer, B., J. Andréasson, S. Fogelberg, H. Johnsson, C.B. Pers, and K. Persson. 2005. Climate change impact on water quality: Model results from southern Sweden. *J Hum Environ* 34(7):559–566.

Armaroli, C., P. Ciavola, L. Perini et al. 2012. Critical storm thresholds for significant morphological changes and damage along the Emilia-Romagna coastline, Italy. *Geomorphology* 143–144:34–51.

Arnell, N.W., S.J. Halliday, R.W. Battarbee, R.A. Skeffington, and A.J. Wade. 2015. The implications of climate change for the water environment in England. *Prog Phys Geog* 39:93–120.

Arshad, M.A., B. Lowery, and B. Grossman, 1996. Physical tests for monitoring soil quality. In: *Methods for Assessing Soil Quality*, SSSA, Madison, WI. Doran, J.W. and A.J. Jones (Eds.). Soil Science Society of America Special Publication Vol. 49, pp. 123–142.

Attandoh, N., S.M. Yidana, A. Abdul-Samed, P.A. Sakyi, B. Banoeng-Yakubo, and P.M. Nude. 2013. Conceptualization of the hydrogeological system of some sedimentary aquifers in Savelugu–Nanton and surrounding areas. *Northern Ghana. Hydrol Process* 27(11):1664–1676.

Atwell, B.J., and B.T. Steer. 1990. The effect of oxygen deficiency on uptake and distribution of nutrients in maize plants. *Plant Soil* 122:1–8.

Auckland Regional Council 2005. Estimating sediment yield. Universal Soil Loss Equation (USLE). Land Fact 8. Auckland, Auckland Regional Council.

Azaroual, M., M. Pettenati, J. Casanova, and N. Rampnoux. 2009. Reactive transport modelling of pollutant transfer through the unsaturated soil zone in the framework of the artificial recharge of an aquifer under seawater intrusion constraints. In: *Proceedings of "REUSE09"*, Brisbane, 22–25 Sept 2008, p. 3.

Azaroual, M.A., L.M. Pettenati, L. André, J. Casanova, and N. Rampnoux. 2008. Reactive transport simulation of the pollutant transfer through the unsaturated soil zone in the framework of an aquifer artificial recharge process. In: *Proceedings of "Water Down Under 2008"*, IBSN 0 858 25735 1, Engineers Australia, Apr 2008, p. 12.

Aziz, K.A., and R.B. Kellogg. 1981. Finite element analysis of a scattering problem. *Math Comput* 37:261–271.

References

Babuška, I., and B.Q. Guo. 1996. The approximation properties of the h-p version of the finite element method. *Comput Method Applied M* 36:319–346.

Baker, G.H., G. Brown, K. Butt, J.P. Curry, and J. Scullion. 2006. Introduced earthworms in agricultural and reclaimed land: Their ecology and influences on soil properties, plant production and other soil biota. *Biol Invasions* 8:1301–1316.

Balba, A.M. 1995. *Management of Problem Soils in Arid Ecosystems.* CRC Press, Boca Raton, Florida. p. 250.

Baldock, J.A., I. Wheeler, N. McKenzie, and A. McBrateny. 2012. Soils and climate change: Potential impacts on carbon stocks and greenhouse gas emissions, and future research for Australian agriculture. *Crop Pasture Sci* 63:269–283.

Bardgett, R.D. 2005. *The Biology of Soil: A Community and Ecosystem Approach.* Oxford University Press, Oxford, UK.

Bardgett, R.D., D.K. Leemans, R. Cook, and P.J. Hobbs. 1997. Seasonality of the soil biota of grazed and ungrazed hill grasslands. *Soil Biol Biochem* 29:1285–1294.

Baric, A., B. Grbec, and D. Bogner. 2008. Potential implications of sea-level rise for Croatia. *J Coastal Research* 24:299–305.

Barry, J.P. 2010. Marine organisms and ecosystems in a high-CO2 ocean and an overview of recommendations from the national research council's committee report on development of an integrated science strategy for ocean acidification monitoring, research, and impacts assessment. Subcommittee on Oceans, Atmosphere, Fisheries, and Coast Guard of the Committee on Commerce. *Science, and Transportation United States Senate.*

Bates, B.C., Z.W. Kundzewicz, S. Wu, and J.P. Palutikof 2008. *Climate Change and Water, Technical Paper IV of the Intergovernmental Panel on Climate Change.* IPCC Secretariat, Geneva. p. 210.

Batlle-Bayer, L. 2010. Changes in organic carbon stocks upon land use conversion in the Brazilian Cerrado: A review. *Agric Ecosyst Environ* 137:47–58.

Beck, A.E. 1981. *Physical Principles of Exploration Methods.* Macmillan, London.

Bekele, E., S. Toze, B. Patterson, and S. Higginson. 2011. Managed aquifer recharge of treated wastewater: Water quality changes resulting from infiltration through the vadose zone. *Water Res* 45(17):5764–5772.

Berner, R.A., and Z. Kothavala. 2001. Geocarb III: A revised model of atmospheric CO_2 over phanerozoic time. *Am J Sci* 301:182–204.

Beutel, M.W. 2006. Inhibition of ammonia release from anoxic profundal sediments in lakes using hypolimnetic oxygenation. *Ecol. Eng.* 28(3):271–279.

Bitton, G., and R.W. Harvey, 1992. Transport of pathogens through soils and aquifers. In: *Environmental Microbiology*, Mitchell, R. (Ed.). Wiley-Liss, Inc, New York, pp. 103–124.

Blanc, E., and R. Reilly. 2015. Climate change impacts on US crops. *Choices* 30(2):1–4.

Blume, H.P. 2002. Some aspects of the history of German soil science. *J Plant Nutr Soil Sci* 165:377–381.

Bobicki, E.R., Q. Liu, Z. Xu, and H. Zeng. 2012. Carbon capture and storage using alkaline industrial wastes. *Prog Energy Combust Sci* 38:302–320.

Bockheim, J.G., and A.N. Gennadiyev. 2000. The role of soil-forming processes in the definition of taxa in soil taxonomy and the world soil reference base. *Geoderma* 95:53–72.

Bollmeyer, C., J.D. Keller, C. Ohlwein et al. 2015. Towards a high-resolution regional reanalysis for the European CORDEX domain. *Q J Roy Meteor Soc* 141:1–15.

Bölük, G.M., and M. Mert. 2014. Fossil and renewable energy consumption, GHGs (greenhouse gases) and economic growth: Evidence from a panel of EU (European Union) countries. *Energy* 74:439–446.

Bonte, M., and G.J.J. Zwolsman. 2010. Climate change induced salinisation of artificial lakes in the Netherlands and consequences for drinking water production. *Water Res.* 44(15):4411–4424.

Boros, G., M. Søndergaard, P. Takács, A. Vári, and I. Tátrai. 2011. Influence of submerged macrophytes, temperature, and nutrient loading on the development of redox potential around the sediment–Water interface in lakes. *Hydrobiologia* 665(1):117–127.

Bot, A., and J. Benites 2005. *The Importance of Soil Organic Matter Key to Drought-resistant Soil and Sustained Food and Production.* FAO Soils Bulletin 80, Food and Agriculture Organization of the United Nations, Rome. p. 78.

Bouraoui, F., G. Vachaud, L.Z. Li, X. Le, H. Treut, and T. Chen. 1999. Evaluation of the impact of climate changes on water storage and groundwater recharge at the watershed scale. *Clim Dyn* 15:153–161.

Brady, N., and R. Weil 2002. *The Nature and Properties of Soils.* 13th ed. Prentice Hall, Upper Saddle River, New Jersey. p. 960.

Bramble, J.H., and J.M. Xu. 1989. Local post-processing technique for improving the accuracy in mixed finite element approximations. *SIAM J Numer Anal* 26:1267–1275.

Bréda, N., A. Granier, F. Barataud, and C. Moyne. 1995. Soil water dynamics in an oak stand. I. Soil moisture, water potentials and water uptake by roots. *Plant Soil* 172:17–27.

Bredehoeft, J.D., and D.L. Norton 1990. *The Role of Fluid in Crustal Processes.* National Academy of Press, Washington, DC, p. 170.

Brekke, L.D., J.E. Kiang, J.R. Olsen et al. 2009. *Climate Change and Water Resources Management, a Federal Perspective.* U.S. Geological Survey Circular 1331, U.S. Geological Survey, Reston.

Brennan, S.T., R.C. Burruss, M.D. Merrill, P.A. Freeman, and L.F. Ruppert. 2010. A probabilistic assessment methodology for the evaluation of geologic carbon dioxide storage: U.S. Geological Survey Open-File Report 2010–1127, p. 31.

Brevik, E.C., 2009. Soil health and productivity. In: *Soils, Plant Growth and Crop Production*, Verheye, W. (Ed.). Encyclopedia of Life Support Systems (EOLSS), Developed under the Auspices of the UNESCO, EOLSS Publishers, Oxford, UK, 2009, p. 106.

Brevik, E.C. 2012. Soils and climate change: Gas fluxes and soil processes. *Soil Horiz* 53:12–23.

Brevik, E.C., 2013. An introduction to soil science basics. In: *Soils and Human Health*, Brevik, E.C. and L.C. Burgess (Eds.). CRC Press, Boca Raton, FL, 2013, pp. 3–28.

Brinkman, R., and H. Brammer, 1990. The influence of a changing climate on soil properties. *Proceedings of the Transactions 14th International Congress of Soil Science*, August 1990, Kyoto, Japan, pp. 283–288.

Briones, M.J.I., M.H. Garnett, and P. Ineson. 2010. Soil biology and warming play a key role in the release of 'old C' from organic soils. *Soil Biol Biochem* 42:960–967.

Brooks, B., T. Valenti, W.T. Valenti et al., 2011. Influence of climate change on reservoir water quality assessment and management. In: *Climate*. Springer, New York, pp. 491–522.

Brouyere, S., G. Carabin, and A. Dassargues. 2004. Climate change impacts on groundwater resources: Modelled deficits in a chalky aquifer, Geer basin, Belgium. *Hydrogeol J* 12:123–134.

References

Brown, E.A., C.H. Wu, D.M. Mickelson, and T.B. Edil. 2005. Factors controlling rates of bluff recession at two sites on Lake Michigan. *J Great Lakes Res* 31:306–321.

Buol, S.W. 2010. Evolution of the text soil genesis and classification. *Soil Survey Horizons* 51:116–117.

Buol, S.W., F.D. Hole, and R.J. McCracken 1973. *Soil Genesis and Classification.* The Iowa State University Press, Ames, Iowa.

Burke, L.K., K. Yumiko, K. Kassem, C. Revenga, M. Spalding, and D. McAllister 2001. *Pilot Analysis of Global Ecosystems: Coastal Ecosystems.* World Resources Institute, Washington, DC, USA. p. 94.

Busari, O., and J. Mutamba. 2014. Groundwater development for localized water supply in South Africa. *Journal of Medical and Bioengineering* 3:4.

Buxton, H.T., and E. Modica. 1992. Patterns and rates of ground-water flow on Long Island, New York. *Ground Water* 30:857–866.

Buxton, H.T., T.E. Reilly, D.W. Pollock, and D.A. Smolensky. 1991. Particle tracking analysis of recharge areas on Lond Island, New York. *Ground Water* 29:63–71.

Caldeira, K., and M.E. Wickett. 2003. Anthropogenic carbon and ocean pH. *Nature* 425(6956):365.

Campbell, G.S. 1974. Simple method for determining unsaturated conductivity from moisture retention data. *Soil Sci* 117:311–314.

Carluer, N., and G. Marsily. 2004. Assessment and modelling of the influence of man-made networks on the hydrology of a small watershed: Implications for fast flow components, water quality and landscape management. *J Hydrol* 285:76–95.

Carter, M.R. 2002. Soil quality for sustainable land management: Organic matter and aggregation interactions that maintain soil functions. *Agron J* 94:38–47.

Carvalho, L., and A. Kirika. 2003. Changes in shallow lake functioning: Response to climate change and nutrient reduction. *Hydrobiologia* 506(1):789–796.

Cave, L., H.E. Beekman, and J. Weaver, 2003. Impact of climate change on groundwater recharge estimation. In: *Groundwater Recharge Estimation in Southern Africa,* Xu, Y. and H.E. Beekman (Eds.). UNESCO, IHP Series No. 64, UNESCO Paris, ISBN 92-9220-000-3, pp. 189–197.

Central Intelligence Agency 2016. *The World Factbook 2016–17: Geography: World: Geographic Overview: Coastline.* Central Intelligence Agency, Washington, DC.

Chafidz, A., S. Al-Zahrani, M.N. Al-Otaibi, C.F. Hoong, T.F. Lai, and M. Prabu. 2014. Portable and integrated solar-driven desalination system using membrane distillation for arid remote areas in Saudi Arabia. *Desalination* 345:36–49.

Chalhoub, M., M. Bernier, Y. Coquet, and M. Philippe. 2017. A simple heat and moisture transfer model to predict ground temperature for shallow ground heat exchangers. *Renew Energ* 103:295–307.

Charman, P.E.V., and B.W. Murphy 2007. *Soils: Their Properties and Management.* Oxford University Press, Melbourne.

Chaskalovic, J. 2008. *Finite Elements Methods for Engineering Sciences.* Springer Verlag. Berlin, Heidelberg.

Chen, C.W., and A. Tang 2011. *Evaluation of Connection Strength of Geogrid to Gabion Wall.* Geotechnical Special Publication No. Vol. 220, pp. 231–238 ASCE.

Chen, M., Y. Chen, S. Dong et al. 2018. Mixed nitrifying bacteria culture under different temperature dropping strategies: Nitrification performance, activity, and community. *Chemosphere* 195:800–809.

Chen, Z., S.E. Grasby, and K.G. Osadetz. 2004. Relation between climate variability and groundwater levels in the upper carbonate aquifer, southern Manitoba, Canada. *J Hydrol* 290:42–62.

Cheney, W., and D. Kincaid 1999. *Numerical Mathematics and Computing*. Brooks/Cole Publishing Co., Pacific Grove, CA, 4th ed., 1999. p. 671.

Cheng, J., Z. Wu, P. Liu, X. Liu, and X. Cai. 2007. AnnAGNPS modeling of agricultural non-point source pollution in the typical watershed of pearl river delta. *J Agro-Environ Sci* 26(3):842–846 (in Chinese).

Childs, E.C., and E.G. Youngs. 2006. The nature of the drain channel as a factor in the design of a land-drainage system. *J Soil Sci* 9:316–331.

Chizmeshya, A.V.G., M.J. McKelvy, D. Gormley et al. 2004 CO_2 mineral carbonation processes in olivine feedstock: Insights from the atomic scale simulation, 29th International Technical Conference on Coal Utilization & Fuel Systems, Coal Technology Association, Clearwater, FL, USA.

Čiamporová-Zaťovičová, Z.L., H.L. Hamerlik, F. Sporka, and P. Bitusik. 2010. Littoral benthic macroinvertebrates of alpine lakes (Tatra Mts) along an altitudinal gradient: A basis for climate change assessment. *Hydrobiologia* 648(1):19–34.

Clapp, R.B., and G.M. Hornberger. 1978. Empirical equations for some soil hydraulic properties. *Water Resour Res* 14:601–604.

Clark, J.R. 1996. *Coastal Zone Management Handbook*. CRC Press.

Clark, L. 1977. The analysis and planning of step drawdown tests. *Q. J. Eng. Geol. Hydrogeol.* 10:125–143.

Clarke, D., and J. Smethurst. 2010. Effects of climate change on cycles of wetting and drying in engineered clay slopes in England. *Q. J. Eng. Geol. Hydrogeol.* 43(4):473–486.

Clarke, G., M. Kernan, A. Marchetto, S. Sorvari, and J. Catalan. 2005. Using diatoms to assess geographical patterns of change in high-altitude European lakes from pre-industrial times to the present day. *Aquat Sci Res Bound* 67(3):224–236.

Cleuvers, M., and T.H. Ratte. 2002. The importance of light intensity in algal tests with coloured substances. *Water Res* 36(9):2173–2178.

Cohen, G., and S. Fauqueux. 2000. Mixed finite elements with mass–lumping for the transient wave equation. *J Comput Acoust* 8:171–188.

Cole, M.A., R.J.R. Elliott, and K. Shimamoto. 2005. Industrial characteristics, environmental regulations and air pollution: An analysis of the UK manufacturing sector. *J Environ Econ Manag* 50:121–143.

Collins, W.J., N. Bellouin, M. Doutriaux-Boucher et al. 2011. Development and evaluation of an earthsystem model – HadGEM2. *Geosci Model Dev* 4:1051–1075.

Connected Waters Initiative 2013. Groundwater levels and aquifer storage. Available at: www.connectedwaters.unsw.edu.au/schools-resources/factsheets/groundwater-levels-and-aquifer-storage.

Connolly, E.L., and E.L. Walker. 2008. Time to pump iron: Iron-deficiency-signaling mechanisms of higher plants. *Curr Opin Plant Biol* 11:530–535.

Corwin, D.L., D.J. Rhoades, and J. Simunek. 2007. Leaching requirement for soil salinity control: Steady-state versus transient models. *Agric Water Manag* 90:165–180.

Cotrufo, M.F., M. Wallenstein, M.C. Boot, K. Denef, and E.A. Paul. 2013. The Microbial Efficiency-Matrix Stabilization (MEMS) framework integrates plant litter decomposition with soil organic matter stabilization: Do labile plant inputs form stable soil organic matter?. *Glob Change Biol* 19:988–995.

Cox, S., T. Crews, and J. Wes. 2013. From genetics and breeding to agronomy to ecology. In: *Perennial Crops for Food Security: Proceedings of the FAO Expert Workshop*, FAO, Rome, pp. 158–168.

Cramer, M.D., H.J. Hawkins, and G.A. Verboom. 2009. The importance of nutritional regulation of plant water flux. *Oecologia* 161:15–24.

References

Crank, J. 1956. *The Mathematics of Diffusion.* Oxford University Press, New York. pp. 12–15.
Cu, O. 1993. Land degradation and the coastal environment of Nigeria. *Catena* 20(3):215–225.
Cubbage, F., G. Balmelli, A. Bussoni et al. 2013. Comparing silvopastoral systems and prospects in eight regions of the world. *Agrofor Syst* 86:303–314.
Czarnes, S., P.D. Hallett, A.G. Bengough, and I.M. Young. 2000. Root and microbial-derived mucilages affect soil structure and water transport. *Eur J Soil Sci* 51:435–443.
Dabbagh, T.A. 2001. The management of desalinated water. *Desalination* 135(1–3):7–23.
Dalla Valle, M., E. Codato, and A. Marcomini. 2007. Climate change influence on POPs distribution and fate: A case study. *Chemosphere* 67(7):1287–1295.
Daly, G.L., and F. Wania. 2005. Organic contaminants in mountains. *Environ Sci Technol* 39(2):385–398.
Damiata, B.N., and T.C. Lee. 2006. Simulated gravitational response to hydraulic testing of unconfined aquifers. *J Hydrol* 318:348–359.
Das, U.C. 1995. Apparent resistivity curves in controlled source electromagnetic sounding directly reflecting true resistivities in a layered earth. *Geophysics* 60:53–60.
Davies, S., H. Mirfenderesk, R. Tomlinson, and S. Szylkarski. 2009. Hydrodynamic, water quality and sediment transport modeling of estuarine and coastal waters on the gold coast Australia. *J Coastal Res* 56:937–941.
Dawson, J.J.C., A.I. Malcolm, J.S. Middlemas, D. Tetzlaff, and C. Soulsby. 2009. Is the composition of dissolved organic carbon changing in upland acidic streams? *Environ Sci Technol* 43(20):7748–7753.
De Deyn, G.B., R.S. Shiel, N.J. Ostleet et al. 2011. Additional carbon sequestration benefits of grassland diversity restoration. *J Appl Ecol* 48:600–608.
de Leeuw, J., M. Njenga, B. Wagner, and M. Iiyama 2014. *Treesilience: An Assessment of the Resilience Provided by Trees in the Dry Lands of Eastern Africa.* ICRAF, Nairobi, Kenya.
De Montety, V., O. Radakovitch, C. Vallet-Coulomb, B. Blavoux, D. Hermitte, and V. Vincent Valles. 2008. Origin of groundwater salinity and hydrogeochemical processes in a confined coastal aquifer: Case of the Rhône delta (Southern France). *Appl Geochem* 23:2337–2349.
Dee, D.P., S.M. Uppala, A.J. Simmons et al. 2011. The ERA-Interim reanalysis: Configuration and performance of the data assimilation system. *Q J Roy Meteor Soc* 137:553–597.
DeJong-Hughes, J.M., J.B. Swan, J.F. Moncrief, and W.B. Voorhees 2001. *Soil Compaction: Causes, Effects and Control (revision).* University of Minnesota Extension Service BU-3115-E. Minnesota.
Delpla, I., V.A. Jung, E. Baures, M. Clement, and O. Thomas. 2009. Impacts of climate change on surface water quality in relation to drinking water production. *Environ. Int.* 35(8):1225–1233.
Dennis, I., and R. Dennis 2013. *Potential Climate Change Impacts on Karoo Aquifers.* WRC Project No KV 308/12; Water Research Commission of South Africa, Pretoria.
Derner, J., L. Joyce, R. Guerrero, and R. Steele, 2015. *Northern Plains Regional Climate Hub Assessment of Climate Change Vulnerability and Adaptation and Mitigation Strategies.* In: Anderson, T. (Ed.). United States Department of Agriculture, Fort Collins, CO, p. 57.
Desortová, B., and P. Punčochář. 2011. Variability of phytoplankton biomass in a lowland river: Response to climate conditions. *Limnol Ecol Manage Inland Waters* 41(3):160–166.

Devia, G.K., B.P. Ganasri, and G.S. Dwarakish. 2015. A review on hydrological models. *Aquatic Procedia* 4:1001–1007.

Dewandel, B., P. Lachassagne, and A. Qatan. 2004. Spatial measurements of stream base flow, a 474 relevant method for aquifer characterization and permeability evaluation. *Application to a 475 Hard Rock Aquifer, the Oman Ophiolite, Hydrol Processes* 18:3391–3400.

Dexter, A.R. 1987. Compression of soil around roots. *Plant and Soil* 97:401–406.

Dijkstra, F.A., S.E. Hobbie, and P.B. Reich. 2006. Soil processes affected by sixteen grassland species grown under different environmental conditions. *Soil Sci Soc Am J* 70:770–777.

Dillon, P. 2005. Future management of aquifer recharge. *Hydrogeol J* 13:313–316.

Dillon, P., and P. Pavelic. 1996. Guidelines on the quality of stormwater and treated wastewater for injection into aquifers for storage and reuse. Urban Water Research Assoc of Aust Research Report N° 109.

Dillon, P., P. Pavelic, S. Toze et al. 2006. Role of aquifer storage in water reuse. *Desalination* 188:123–134.

Dingman, S.L. 2002. *Physical Hydrology. Prentice-Hall*. Upper Saddle River, New Jersey, 2nd ed., p. 646.

Dinnes, D.L. 2004. *Assessment of Practices to Reduce Nitrogen and Potassium Non-point Source Pollution of Iowa's Surface Waters*. Iowa Dept. of National resources, Des Moines, LA.

DOE. 1993. Assessment of the vulnerability of coastal areas to sea level rise and other effects of global climate change, Pilot Study Bangladesh, Report prepared by Department of Environment, Govt. of Bangladesh, Dhaka.

Döll, P. 2009. *Vulnerability to The Impact of Climate Change on Renewable Groundwater Resources: A Global-Scale Assessment*. IOP Publishing Ltd, Bristol BS1 6HG. Royaume-Uni.

Domenico, P.A., and F.W. Schwartz 1990. *Physical and Chemical Hydrogeology*. John Wiley & Sons, New York, NY.

Doocy, S., A. Daniels, S. Murray, and T. Kirsch. 2013. The human impact of floods: A historical review of events 1980–2009 and systematic literature review. *PLOS Current Disasters* 5:1–29.

Dragoni, W., and B.S. Sukhija, 2008. Climate change and groundwater – A short review. In: *Climate Change and Groundwater*, Dragoni, W. and B.S. Sikhija (Eds.). Geological Society, Special Publications, London, Vol. 288: pp. 1–12.

Drew, M., 1988. Effects of flooding and oxygen deficiency on the plant mineral nutrition. In: *Advances in Plant Nutrition*, Tinker, E., P.B. Tinker and A. Lauchli (Eds.). Greenwood Publishing Group, Westport, Connecticut, pp. 115–159.

Drewry, J., L. Newham, and W.F.B. Croke. 2009. Suspended sediment, nitrogen and phosphorus concentrations and exports during storm-events to the Tuross estuary, Australia. *J Environ Manage* 90(2):879–887.

Ducharne, A., C. Baubion, N. Beaudoin et al. 2007. Long term prospective of the Seine River system: Confronting climatic and direct anthropogenic changes. *Sci Total Environ* 375(1–3):292–311.

Ducharne, A., F. Habets, C. Pagé et al. 2010. Climate change impacts on water resources and hydrological extremes in northern France. Paper presented at the XVIII International Conference on Water Resources, Barcelona (Spain).

Dunn, S.M., and R. Mackay. 1996. Modelling the hydrological impacts of open ditch drainage. *J Hydrol* 179:37–66.

References

Düster, A., H. Broeker, and E. Rank. 2001. The p–Version of the finite element method for three–Dimensional curved thin walled structures. *International Journal for Numerical Methods in Engineering* 52:673–703.

Earman, E., and M. Dettinger. 2011. Potential impacts of climate change on groundwater resources: A global review. *J Water Clim Change* 2(4):213–229.

Egholm, D.L., S.B. Nielsen, V.K. Pedersen, and J.E. Lesemann. 2009. Glacial effects limiting mountain height. *Nature* 460(7257):884–887.

Eimers, M.C., J.P. Dillon, and L.S. Schiff. 2004. Sulphate flux from an upland forested catchment in south-central Ontario, Canada. *Water Air Soil Pollut* 152:3–21.

Eimers, M.C., S.A. Watmough, J.M. Buttle, and P.J. Dillon. 2008. Examination of the potential relationship between droughts, sulphate and dissolved organic carbon at a wetland-draining stream. *Global Change Biol* 14(4):938–948.

Elsdon, T.S., M.A.N.B. DeBruin, J.N. Diepen, and M.B. Gillandersa. 2009. Extensive drought negates human influence on nutrients and water quality in estuaries. *Sci Total Environ* 407(8):3033–3043.

Engman, E.T., and N. Chauhan. 1995. Status of microwave soil moisture measurements with remote sensing. *Remote Sens Environ* 51:189–198.

Enríquez, A.R., M. Marcos, A. Álvarez-Ellacuría, A. Orfila, and D. Gomis. 2017. Changes in beach shoreline due to sea level rise and waves under climate change scenarios: Application to the Balearic Islands (western Mediterranean). *Nat Hazard Earth Sys* 17:1075–1089.

Eurosion. 2004. Living with Coastal Erosion in Europe: Sediment and Space for Sustainability. Part IV, a guide to coastal erosion management practices in Europe: lessons learned. EUCC. France. p. 27.

Evans, C., T.D. Monteith, and M.D. Cooper. 2005. Long-term increases in surface water dissolved organic carbon: Observations, possible causes and environmental impacts. *Environ Pollut* 137(1):55–71.

Ewen, J., G. Parkin, and P.E. O'Connell. 2000. Shetran: Distributed river basin flow and transport modeling system. *J Hydrol Eng* 5:250–258.

Falgàs, E., J. Ledo, B. Benjumea et al. 2011. Integrating hydrogeological and geophysical methods for the characterization of a deltaic aquifer system. *Surv Geophys* 32(6):857–873.

Fang, H.Y., and J. Daniels 2005. *Introductory Geotechnical Engineering: An Environmental Perspective*. Taylor & Francis, London.

FAO 2007. *Adaptation to Climate Change in Agriculture, Forestry, and Fisheries: Perspective, Framework and Priorities*. Food and Agriculture Organization of the United Nations (FAO), Interdepartmental Working Group on Climate Change. 32, Rome, Italy.

Faour, G., A. Fayad, and M. Mhawej. 2013. GIS-based approach to the assessment of coastal vulnerability to sea level rise: Case study on the eastern mediterranean. *J Surv Map Eng* 12(3):41–48.

Farhadzadeh, A., N. Kobayashi, and B. Gravens. 2012. Effect of breaking waves and external current on longshore sediment transport. *J Waterw Port C-asce* 138:256–260.

Feagin, R.A., D.J. Sherman, and W.E. Grant. 2005. Coastal erosion, global sea-level rise, and the loss of sand dune plant habitats. *Front Ecol Environ* 3:359–364.

Fedorova, E., E. Sviridova, K. Marusin, and A. Khabidov. 2010. Remote and cartographical techniques for estimation of coastal erosion rate in seas and inland water bodies. In: *Proceedings of the International Conference on Dynamics of Coastal*

Zone of Non-tidal Seas, 26–30 June 2010, Baltiysk (Kaliningrad Oblast, Russia), pp. 82–84.

Feely, R.A., C.L. Sabine, K. Lee et al. 2004. Impact of anthropogenic CO_2 on the $CaCO_3$ system in the oceans. *Science* 305(5682):362–366.

Feitelson, E., and G. Rosenthal. 2012. Desalination, space and power: The ramifications of Israel's changing water geography. *Geoforum* 43(2):272–284.

Fenton, G., and K. Helyar, 2007. Soil acidification. In: *Soils: Their Properties and Management*, Charman, P.E.V. and B.W. Murphy (Eds.). 3rd ed. Oxford University Press, Melbourne, pp. 224–237.

Ferguson, G., and T. Gleeson. 2012. Vulnerability of coastal aquifers to groundwater use and climate change. *Nature Climate Change* 2:342–345.

Ferreira, O., P. Ciavola, C. Armaroli et al. 2009. Coastal storm risk assessment in Europe: Examples from 9 study sites. *J Coast Res* 56:1632–1636.

Fetter, C.W. 2001. *Applied Hydrogeology*. 4th ed. Prentice Hall, New Jersey.

Ficklin, D.L., Y. Luo, E. Luedeling, and M. Zhang. 2009. Climate change sensitivity assessment of a highly agricultural watershed using SWAT. *J Hydrol* 374(1):16–29.

Fischer, E.M., S. Seneviratne, D. Lüthi, and C. Schär. 2007. Contribution of land–Atmosphere coupling to recent European summer heat waves. *Geophys Res Lett* 34:L06707.

Fischer, J.M., H.M. Olson, E.C. Williamson et al. 2011. Implications of climate change for Daphnia in alpine lakes: Predictions from long-term dynamics, spatial distribution, and a short-term experiment. *Hydrobiologia* 676:263–277.

FitzGerald, D.M., M.S. Fenster, B.A. Argow, and I.V. Buynevich. 2008. Coastal impacts due to sea level rise. *Annu Rev Earth Pl Sc* 36:601–647.

Florek, W., J. Kaczmarzyk, M. Majewski, and I.J. Olszak. 2008. Lithological and extreme event control of changes in cliff morphology in the Ustka region. *Landform Analysis* 7:53–68.

Fontaine, C.M., M. Mokrech, and M.D.A. Rounsevell. 2015. Land use dynamics and coastal management. *Adv Clim Change Res* 49:125–146.

Forbes, K.A., W.S. Kienzle, A.C. Coburn, M.J. Byrne, and J. Rasmussen. 2011. Simulating the hydrological response to predicted climate change on a watershed in southern Alberta. *Canada, Climatic Change* 105(3–4):1–22.

Forster-Carneiro, T., M. Berni, I. Dorileo, and M. Rostagno. 2013. Biorefinery study of availability of agriculture residues and wastes for integrated biorefineries in Brazil. *Resour Conserv Recycl* 77:78–88.

Foster, S., J. Chilton, G.J. Nijsten, and A. Richts. 2013. Groundwater, a global focus on the 'local resource'. *Curr Opin Environ Sustain* 5(6):685–695.

Freeze, R.A., and J.A. Cherry 1979. *Groundwater*. Prentice Hall, New Jersey. p. 604.

Freidman, D., M. Hubbs, A. Tugel, C. Seybold, and M. Sucik 2001. *Guidelines for Soil Quality Assessment in Conservation Planning*. US Government Printing Office, Washington, DC.

Fulweiler, R.W., and W.S. Nixon. 2009. Responses of benthicpelagic coupling to climate change in a temperate estuary. *Eutrophication in Coastal Ecosystems* 207:147–156.

Gallart, F., P. Llorens, and J. Latron. 1994. Studying the role of old agricultural terraces on runoff generation in a small Mediterranean mountainous basin. *J Hydrol* 159:291–303.

Gantzer, P.A., D.L. Bryant, and C.J. Little. 2009. Controlling soluble iron and manganese in a water-supply reservoir using hypolimnetic oxygenation. *Water Res* 43(5):1285–1294.

Ge, X., T. Li, S. Zhang, and M. Peng. 2010. What causes the extremely heavy rainfall in Taiwan during Typhoon Morakot (2009)? *Atmospheric Science Letters* 11(1):46–50.

References

Gee, G.W., and J.W. Bauder, 1986. Particle size analysis. In: *Methods of Soil Analysis, Part I*, 2nd ed. A. Klute (Ed.). Amer Soc Agron, Madison, pp. 383–411.

Gelybó, G., E. Tóth, C. Farkas, Á. Horel, I. Kása, and Z. Bakacsi. 2018. Potential impacts of climate change on soil properties. *AgroChemistry and Soil Science* 67(1):121–141.

Gerdemann, S.J., D.C. Dahlin, and W.K. O'Connor. 2002. Carbon dioxide sequestration by aqueous mineral carbonation of magnesium silicate minerals, 6th International Conference on Greenhouse Gas Control Technologies, Kyoto, Japan.

Gerrish, J. 2004. *Management-intensive Grazing: The Grassroots of Grass Farming*. Green Park Press, Ridgeland, MS.

Gill, R.A., H.W. Polley, H.B. Johnson, L.J. Anderson, H. Maherali, and R.B. Jackson. 2002. Nonlinear grassland responses to past and future atmospheric CO2. *Nature* 417:279–282.

Gill, S.L., F.C. Spurlock, K.S. Goh, and C. Ganapathy. 2008. Vegetated ditches as a management practice in irrigated alfalfa. *Environ Monit Assess* 144:261–267.

Gillanders, B.M., T.S. Elsdon, I.A. Halliday, G.P. Jenkins, J.B. Robins, and F.J. Valesini. 2011. Potential effects of climate change on Australian estuaries and fish utilizing estuaries: A review. *Mar and Freshwater Res* 62(9):1115–1131.

Gillot, J.E. 1981. Book review of: Soil genesis and classification, 2nd ed. *Canadian Geotechnical Journal* 18:609.

Giuntoli, I., B. Renard, J.P. Vidal, and A. Bard. 2013. Low flows in France and their relationship to largescale climate indices. *J Hydrol* 482:105–118.

Glassley, W.E., J.J. Nitao, C.W. Grant, J.W. Johnson, C.I. Steefel, and J.R. Kercher. 2003. The impact of climate change on vadose zone pore waters and its implication for long-term monitoring. *Comput Geosci* 29(3):399–411.

Gleeson, T., Y. Wada, M.F.P. Bierkens, and L.P.H. van Beek. 2012. Water balance of global aquifers revealed by groundwater footprint. *Nature* 488:197–200.

Goldberg, P., Z.Y. Chen, W. O'Connor, R. Walters, and H. Ziock. 2001. CO_2 mineral sequestration situation in US. *Journal of Energy and Environmental Research* 1:117.

Gómez, R., I.M. Arce, J.J. Sonchez, and M.M. Sonchez-Montoya. 2012. The effects of drying on sediment nitrogen content in a Mediterranean intermittent stream: A microcosms study. *Hydrobiologia* 679(1):43–59.

Goosse, H., P.Y. Barriat, W. Lefebvre, M.F. Loutre, and V. Zunz. 2010. Introduction to climate dynamics and climate modeling. Online textbook available at www.climate.be/textbook.

Gopinath, G., and P. Seralathan. 2005. Rapid erosion of the coast of Sagar island, West Bengal – India. *Environ Geol* 48:1058–1067.

Gourlay, T. 2010. Full-scale boat wake and wind wave trials on the Swan River. Centre for Marine Science and Technology, Curtin University research report No: 2010–06.

Gourlay, T. 2011. Notes on shoreline erosion due to boat wakes and wind waves. Centre for Marine Science and Technology, Curtin University research report No: 2011–16.

Govers, G., J. Poesen, D. Goossens, and B.T. Christensen, 2004. Soil erosion: Processes, damages and countermeasures. In: *Managing Soil Quality: Challenges in Modern Agriculture*, Schjonning, P. and S. Elmholt (Eds.). CABI Publ., Wallingford, UK, pp. 199–217.

Grandy, A.S., and J.C. Neff. 2008. Molecular C dynamics downstream: The biochemical decomposition sequence and its impact on soil organic matter structure and function. *Sci. Total Environ* 404:297–307.

Gray, J.M., T.F.A. Bishop, and X. Yang. 2015. Pragmatic models for prediction and digital mapping of soil properties in eastern Australia. *Soil Res* 53:24–42.
Greacen, E.L. 1981. *Soil Water Assessment by Neutron Methods.* C.S.I.R.O, Australia.
Green, T.R., B.C. Bates, S.P. Charles, and P.M. Fleming. 2007. Physically based simulation of potential effects of carbon dioxide: Altered climates on groundwater recharge. *Vadose Zone J* 6(3):597–609.
Green, T.R., M. Taniguchi, H. Kooi et al. 2011. Beneath the surface of global change: Impacts of climate change on groundwater. *J Hydrol* 405(3–4):532–560.
Greenland, D.J. 1997. Inaugural Russell-Memorial-lecture-soil conditions and plant growth. *Soil Use and Manage* 13:169–177.
Groothuizen, A.G.M. 2008. World development and the importance of dredging. *PIANC Magazine*, N° 130.
Gudadhe, N., M.B. Dhonde, and N.A. Hirwe. 2015. Effect of integrated nutrient management on soil properties under cotton-chickpea cropping sequence in vertisols of Deccan plateau of India. *Indian J Agric Res* 49(3):207–214.
Gurdak, J.J., R.T. Hanson, P.B. McMahon, B.W. Bruce, J.E. McCray, G.D. Thyne, and R.C. Reedy. 2007. Climate variability controls on unsaturated water and chemical movement, high plains aquifer, USA. *Vadose Zone J* 6(3):533–547.
Häder, D.P., D.H. Kumar, C.R. Smith, and C.R. Worrest. 2007. Effects of solar UV radiation on aquatic ecosystems and interactions with climate change. *Photochem Sci* 6(3):267–285.
Haeber, R.H., and D.H. Waller. 1987. Water quality of rainwater collection systems in the Eastern Caribbean. Third international rainwater catchment systems conference proceedings.
Hairer, E., and G. Wanner 2010. *Solving Ordinary Differential Equations I, Stiff and Differential-Algebraic Problems.* Springer Verlag, Berlin.
Haj-Amor, Z., H. Ritzema, H. Hashemi, and S. Bouri. 2018. Surface irrigation performance of date palms under water scarcity in arid irrigated lands. *Arab J Geosci* 11:27.
Hamilton, S.K. 2010. Biogeochemical implications of climate change for tropical rivers and floodplains. *Hydrobiologia* 657(1):19–35.
Hammer, M.J., and G. Elser. 1980. Control of ground-water salinity, Orange County, California. *Ground Water* 18(6):536–540.
Hammond, D., and A. Pryce. 2007. Climate change impacts and water temperature.
Han, H., D.J. Allan, and D. Scavia. 2009. Influence of climate and human activities on the relationship between watershed nitrogen input and river export. *Environ Sci Technol* 43(6):1916–1922.
Hari, R.E., M.D. Livingstone, R. Siber, P. Burkhardt-Holm, and H. Guttinger. 2006. Consequences of climatic change for water temperature and brown trout populations in Alpine rivers and streams. *Global Change Biol* 12(1):10–26.
Hartley, I.P., D.W. Hopkins, M.H. Garnett, M. Sommerkorn, and P.A. Wookey. 2008. Soil microbial respiration in arctic soil does not acclimate to temperature. *Ecol Lett* 11:1092–1100.
Hartmann, J., A.J. West, P. Renforth et al. 2013. Enhanced chemical weathering as a geoengineering strategy to reduce atmospheric carbon dioxide, supply nutrients, and mitigate ocean acidification. *Rev Geophys* 51:113–149.
Harvey, C.A., M. Chacón, C.I. Donatti et al. 2014. Climate-smart landscapes: Opportunities and challenges for integrating adaptation and mitigation in tropical agriculture. *Conserv Lett* 7(2):77–90.

Hassink, J. 1994. Effects of soil texture and grassland management on soil organic C and N and rates of C and N mineralization. *Soil Biol Biochem* 26:1221–1231.

Hearn, G.J., and T. Hunt. 2011. Slope and road drainage. *Geological Society, London, Engineering Geology Special Publications* 24:231–242.

Heath, J., E. Ayres, M. Possell et al. 2005. Rising atmospheric CO2 reduces soil carbon sequestration. *Science* 309:1711–1713.

Hell, R., and H. Hillebrand. 2001. Plant concepts for mineral acquisition and allocation. *Curr Opin Biotech* 12:161–168.

Hernádi, H., C. Farkas, A. Makó, and F. Máté. 2009. Climate sensitivity of soil water regime of different hungarian chernozem soil subtypes. *Biologia* 64:624–628.

Herzog, H., B. Eliasson, and O. Kaarstad. 2000. Capturing greenhouse gases. *Scientific American* 282(2):72–77.

Hillel, D. 1973. *Soil and Physical Principles and Processes.* (3rd ed.). Academic Press. Cambridge, MA, p. 248.

Hillel, D. 1998. *Environmental Soil Physics.* Academic Press, San Diego, CA.

Hilscherova, K., L. Dusek, V. Kubik, P. Cupr, J. Hofman, J. Klanova, and I. Holoubek. 2007. Redistribution of organic pollutants in river sediments and alluvial soils related to major floods. *J Soils Sed* 7(3):167–177.

Hirschi, M., S.I. Seneviratne, T. Corti et al. 2011. Observational evidence for soil-moisture impact on hot extremes in southeastern Europe. *Nat Geosci* 4:17–21.

Hodgson, J., and A.W. Illius (Eds.). 1996. *The Ecology and Management of Grazing Systems.* CAB International, Wallingford, UK.

Hodny, J.W. 1992. Climate change and water resources within the Delaware River and Delmarva Regions. Master's thesis, Department of Geography, University of Delaware, Newark.

Hoekstra, P., and M. Blohm, 1990. Case histories of time-domain electromagnetic soundings in environmental geophysics. In: *Geotechnical and Environmental Geophysics, II*, WARD, S.H. (Ed.). Society of Exploration Geophysicists, Tulsa, pp. 1–15.

Hoffman, J.D., and S. Frankel 2001. *Numerical Methods for Engineers and Scientists.* CRC Press, Boca Raton.

Holland, M., D.A. Bailey, B.P. Briegleb, B. Light, and E. Hunke. 2012. Improved sea ice shortwave radiation physics in CCSM4: The impact of melt ponds and aerosols on Arctic sea ice. *J Clim* 25:1413–1430.

Holman, I.P. 2006. Climate change impacts on groundwater recharge: Uncertainty, shortcomings, and the way forward?. *Hydrogeol J* 14:637–647.

Holman, R.A. 1986. Extreme value statistics for wave run-up on a natural beach. *Coast Eng* 9:527–544.

Holsten, A., T. Vetter, K. Vohland, and V. Krysanova. 2009. Impact of climate change on soil moisture dynamics in Brandenburg with a focus on nature conservation areas. *Ecol Model* 220:2076–2087.

Holthuijsen, L.H., N. Booij, and T.H.C. Herbers. 1989. A prediction model for stationary, short-crested waves in shallow water with ambient currents. *Coast Eng* 13:23–54.

Horel, Á., E. Tóth, G. Gelybó, I. Kása, Z. Bakacsi, and C. Farkas. 2015. Effects of land use and management on soil hydraulic properties. *Open Geosci* 7(1):742–754.

Horvat, C., D.R. Jones, S. Iams, D. Schroeder, D. Flocco, and D. Feltham. 2017. The frequency and extent of sub-ice phytoplankton blooms in the Arctic Ocean. *Sci Adv* 3(3):e1601191.

Hoseini, Y., and M. Albaji. 2016. Calibration of gypsum blocks for measuring saline soils moisture. *Commun Soil Sci Plant Anal* 47(22):2528–2537.

Hossain, M.S., and C.K. Lin. 2001. Land use zoning for integrated coastal zone management: Remote sensing, GIS and PRA approach in Cox's Bazar coast, Bangladesh, ITCZM Monograph No. 3, Asian Institute of Technology, Thailand.

Howarth, R.W., P.D. Swaney, J.T. Butler, and R. Marino. 2007. Rapid communication: Climatic control on eutrophication of the Hudson River Estuary. *Ecosystems* 3(2):210–215.

Howden, S.M., J.F. Soussana, F.N. Tubiello, N. Chhetri, M. Dunlop, and H. Meinke. 2007. Adapting agriculture to climate change. *Proceedings of the National Academy of Sciences* 104(50):19691–19696.

Hrdinka, T., O. Nocivký, E. Hanslík, and M. Rieder. 2012. Possible impacts of floods and droughts on water quality. *J Hydro-environ Res* 6(2):145–150.

Hruska, J., P. Kram, H.W. McDowell, and F. Oulehle. 2009. Increased dissolved organic carbon (DOC) in Central European Streams is driven by reductions in ionic strength rather than climate change or decreasing acidity. *Environ Sci Technol* 43(12):4320–4326.

Hu, S., C. Liu, H. Zheng, Z. Wang, and J. Yu. 2012. Assessing the impacts of climate variability and human activities on streamflow in the water source area of Baiyangdian Lake. *J Geogr Sci* 22(5):895–905.

Huang, G., J. Sun, Y. Zhang, Z. Chen, and F. Liu. 2012. Impact of anthropogenic and natural processes on the evolution of groundwater chemistry in a rapidly urbanized coastal area, South China. *Sci Total Environ* 463–464:209–221.

Hughes, T.J.R. 1987. *The Finite Element Method.* Prentice Hall, New Jersey.

Hunke, E.C., and J.K. Dukowicz. 1997. An elastic–viscous–plastic model for sea ice dynamics. *J Phys Oceanogr* 27(9):1849–1867.

Hunter, K.S., Y.F. Wang, and P. Van Cappellen. 1998. Kinetic modeling of microbially-driven redox chemistry of subsurface environments: Coupling transport, microbial metabolism and geochemistry. *J Hydrol* 209:53–80.

Huntley, D. 1986. Relations between permeability and electrical resistivity in granular aquifers. *Ground Water* 24:466–474.

Hupet, F., M.C.T. Trought, M. Greven, S.R. Green, and B.E. Clothier. 2005. Data requirements for identifying macroscopic water stress parameters: A study on grapevines. *Water Resour Res* 41:W06008.

Huq, S., S.I. Ali, and A.A. Rahman. 1995. Sea-level rise and Bangladesh: A preliminary analysis. *J Coast Res* 14:44–53.

Hyndman, D.W., and S.M. Gorelick. 1996. Estimating lithologic and transport properties in three dimensions using seismic and tracer data, the Kesterson aquifer. *Water Resour Res* 32(9):2659–2670.

Idoko, O.M. 2010. Seasonal variation in iron in rural groundwater of Benue State, Middle Belt Nigeria. *Pak J Nutr* 9(9):892–895.

Ingham, R.E., J.A. Trofymow, E.R. Ingham, and D.C. Coleman. 1985. Interactions of bacteria, fungi, and their nematode grazers: Effects on nutrient cycling and plant growth. *Ecol Monogr* 55:119–140.

IPCC. 2007. Summary for policymakers. In Climate Change 2007: The Physical Science Basis; Contribution of Working Group I to the Fourth Assessment Report of the Intergovernmental Panel on Climate Change. Cambridge University Press: Cambridge, UK, pp. 1–18.

IPCC. 2013. Summary for policymakers. In Climate Change 2013: The Physical Science Basis. Contribution of Working Group to the Fifth Assessment Report of the Intergovernmental Panel on Climate Change. Cambridge University Press, Cambridge, UK and New York, NY, USA.

References

IPCC. 2014. Climate change 2014: Synthesis report. Contribution of working groups I, II and III to the fifth assessment report of the intergovernmental panel on climate change. Geneva, p. 151.

Islam, K.S., X. Xue, and M.M. Rahman. 2009. Successful Integrated Coastal Zone Management (ICZM) program model of a developing country (Xiamen, China) – implementation in Bangladesh Perspective. *J Wetlands Ecol* 2(1):1–2.

Izaurralde, R.C., and C.W. Rice, 2006. Methods and tools for designing a pilot soil carbon sequestration project. In: *Carbon Sequestration in Soils of Latin America*, Lal, R., C.C. Cerri, M. Bernoux, J. Etchevers and E. Cerri (Eds.). Haworth Press, Binghamton, NY, pp. 457–476.

Jackson, L.J., L.T. Lauridsen, M. Søndergaard, and E. Jeppesen. 2007. A comparison of shallow Danish and Canadian lakes and implications of climate change. *Freshw Biol* 52(9):1782–1792.

Jacobson, M.Z. 2005. Studying ocean acidification with conservative, stable numerical schemes for nonequilibrium air–ocean exchange and ocean equilibrium chemistry. *Journal of Geophysical Research: Atmospheres* 110:D07302.

Jakeman, A.J., and G.M. Hornberger. 1993. How much complexity is warranted in a rainfall-runoff model? *Water Resour Res* 29:2637–2649.

Jiang, X., X. Jin, Y. Yao, L. Li, and F. Wu. 2008. Effects of biological activity, light, temperature and oxygen on phosphorus release processes at the sediment and water interface of Taihu Lake, China. *Water Res* 42(8–9):2251–2259.

Jing, L., and B. Chen. 2011. Field investigation and hydrological modelling of a subarctic wetland the Deer River Watershed. *J Environ. Inf* 17(1):36–45.

Joehnk, K.D., J. Huisman, J. Sharples, B. Sommeijer, M.P. Visser, and M.J. Stroom. 2008. Summer heatwaves promote blooms of harmful cyanobacteria. *Global Change Biol* 14(3):495–512.

Johnson, J.M.F., R.R. Allmaras, and D.C. Reicosky. 2006. Estimating source carbon from crop residues, roots and rhizodeposits using the national grain-yield database. *Agron J* 98:622–636.

Johnson, J.S., L.A. Baker, and P. Fox. 1999. Geochemical transformations during artificial groundwater recharge: Soil-water interactions of inorganic constituents. *Water Res* 33:196–206.

Johnson, J.W., J.J. Nitao, C.I. Steefel, and K.G. Knauss. 2001. Reactive transport modeling of geologic CO_2 sequestration in saline aquifers: The influence of intra-aquifer shales and the relative effectiveness of structural solubility, and mineral trapping during prograde and retrograde sequestration. In: *Proceedings of the First National Conference on Carbon Sequestration*, Washington D.C., May 14–17, 2001.

Jones, D.L., C. Nguyen, and R.D. Finlay. 2009. Carbon flow in the rhizosphere: Carbon trading at the soil-root interface. *Plant Soil* 321:5–33.

Joss, A., E. Keller, A.C. Alder et al. 2005. Removal of pharmaceuticals and fragrances in biological wastewater treatment. *Water Res* 39:3139–3152.

Kaminsky, G.M., P. Ruggiero, G. Gelfenbaum, and C. Peterson. 1997. Long term coastal evolution and regional dynamics of a US Pacific Northwest littoral cell, *Proceedings of Coastal Dynamics '97, the International Conference on Coastal Research through Large Scale Experiments*, Plymouth, UK, June 23–27.

Kamphuis, J.W. 2002. *Introduction to Coastal Engineering and Management. Advance Series in Ocean Engineering*. World Scientific. Singapore, p. 437.

Kandaris, P.M. 1999. Use of gabions for localized slope stabilization in difficult terrain. The 37th U.S. Symposium on Rock Mechanics (USRMS), pp. 1221–1227, Vail, Colorado.

Karashian, O., and C. Makridakis. 1998. A space–time finite element method for the nonlinear Schrodinger equation: The discontinuous Galerkin method. *Mathematics of Computation* 67:479–499.

Karmakar, R., I. Das, D. Dutta, and A. Rakshit. 2016. Potential effects of climate changes on soil properties: A review. *Sci Int* 4(2):51–73.

Kaste, Ø., R. Wright, J.L. Barkved et al. 2006. Linked models to assess the impacts of climate change on nitrogen in a Norwegian river basin and fjord system. *Sci Total Environ* 365(1–3):200–222.

Kaushal, S.S., M.P. Groffman, E.L. Band et al. 2008. Interaction between urbanization and climate variability amplifies watershed nitrate export in Maryland. *Environ Sci Technol* 42(16):5872–5878.

Kelly, W.E. 1977. Geoelectrical sounding for estimating aquifer hydraulic conductivity. *Ground Water* 15:420–424.

Khan, S., S. Mushtaq, M.A. Hanjra, and J. Schaeffer. 2008. Estimating potential costs and gains from an aquifer storage and recovery program in Australia. *Agric Water Manag* 95(4):477–488.

Kinnell, P.I.A. 2005. Raindrop-impact-induced erosion processes and prediction: A review. *Hydrol. Process* 19:2815–2844.

Kirilova, E., O. Heiri, P. Bluszcz, B. Zolitschka, and F.A. Lotter. 2011. Climate-driven shifts in diatom assemblages recorded in annually laminated sediments of Sacrower See (NE Germany). *Aquat Sci Res Bound* 73(2):201–210.

Kirk, J.L., M.C.D. Muir, J. Antoniades et al. 2011. Climate change and mercury accumulation in Canadian high and subarctic lakes. *Environ Sci Technol* 45(3):964–970.

Kirkham, D., and W.L. Powers 1972. *Advanced Soil Physics*. Wiley-Interscience, A division of John Wiley & Sons, New York, NY.

Kirkham, M.B. 1994. Streamlines for diffusive flow in vertical and Q surface tillage: A model study. *Soil Sci Soc Am J* 58:85–93.

Kloppmann, W., A. Aharoni, H. Chikurel et al., 2012. Use of groundwater models for prediction and optimization of the behavior of MAR sites. In: *Water Reclamation Technologies for Safe Managed Aquifer Recharge*, Christian, K., W. Thomas and D. Peter (Eds.). IWA Publishing, London, pp. 311–349.

Knecht, R.W., and J. Archer. 1993. Integration in the US coastal management program. *Ocean Coast Manage* 21:183–200.

Knowles, N., and R.D. Cayan. 2002. Potential effects of global warming on the Sacramento/San Joaquin watershed and the San Francisco estuary. *Geophys Res Lett* 29 (18):38-1–38-4.

Kock Rasmussen, E., O. Svenstrup Petersen, R.J. Thompson, J.R. Flower, and H. M. Ahmed. 2009. Hydrodynamic-ecological model analyses of the water quality of Lake Manzala (Nile Delta, Northern Egypt). *Hydrobiologia* 622(1):195–220.

Köhler, P., J.F. Abrams, C. Völker, J. Hauck, and D.A. Wolf-Gladrow. 2013. Geoengineering impact of open ocean dissolution of olivine on atmospheric CO_2, surface ocean pH and marine biology. *Environ Res Lett* 8:014009.

Komar, P.D. 2000. Coastal erosion: Underlying factors and human impacts. *Shore and Beach* 68(1):3–16.

Komatsu, E., T. Fukushima, and H. Harasawad. 2007. A modeling approach to forecast the effect of long-term climate change on lake water quality. *Ecol Model* 209(2-4):351–366.

Koster, R.D., M.J. Suarez, W. Higgins, and H.M. Van Den Dool. 2003. Observational evidence that soil moisture variations affect precipitation. *Geophys Res Lett* 30(5):1241.

References

Kresic, N. 1997. *Quantitative Solutions in Hydrogeology and Groundwater Modeling.* Lewis Publishers, New York. p. 461.

Krupavathi, K., and R.B. Movva. 2016. Sea water intrusion into coastal aquifers: Concepts, methods and adoptable control practices. *Int J Agric Eng* 2(9):213–221.

Kruseman, G., and N. de Ridder 1990. *Analysis and Evaluation of Pumping Test Data.* 2nd ed. Publication 47, International Institute for Land Reclamation and Improvement, Wageningen, Netherlands.

Łabuz, T.A., and H. Kowalewska-Kalkowska. 2011. Coastal erosion caused by the heavy storm surge of November 2004 in the southern Baltic Sea. *Clim Res* 48:93–101.

Lackner, K.S., D.P. Butt, and C.H. Wendt. 1997. Progress on binding CO2 in mineral substrates. *Energ Convers Manage* 38:S259-S264.

Lackner, K.S., C.H. Wendt, D.P. Butt, E.L. Joyce, and D.H. Sharp. 1995. Carbon dioxide disposal in carbonate minerals. *Energy* 20(11):1153–1170.

Lal, R. 1999. Soil management and restoration for C sequestration to mitigate the accelerated greenhouse effect. *Progress in Environmental Science* 1:307–326.

Lal, R., 2001. The potential of soil carbon sequestration in forest ecosystem to mitigate the greenhouse effect. In: *Soil Carbon Sequestration and the Greenhouse Effect*, R. Lal (Ed.). Soil Science Society of America Special Publication, Vol. 57. Madison, WI, pp. 151–184.

Lal, R. 2002. The potential of soils of the tropics to sequester carbon and mitigate the greenhouse effect. *Adv Agron* 74:155–192.

Lancelot, C., E. Hannon, S. Becquevort, C. Veth, and H.J.W. De Baar. 2000. Modeling phytoplankton blooms and carbon export production in the Southern Ocean: Dominant controls by light and iron in the Atlantic sector in Austral spring 1992. *Deep-Sea Research Part I-Oceanographic Research Papers* 47(9):1621–1662.

Landes, A.L., A.,.L. Aquilina, J. De Ridder, L. Longuevergne, C. Page, and P. Goderniaux. 2014. Investigating the respective impacts of groundwater exploitation and climate change on wetland extension over 150 years. *J Hydrol* 509:367–378.

Lapworth, D.J., N. Baran, M.E. Stuart, and R.S. Ward. 2012. Emerging organic contaminants in groundwater: A review of sources, fate and occurrence. *Environ Pollut* 163:287–303.

Lashkarripour, G.R. 2003. An Investigation of Groundwater Condition by Geoelectrical Resistivity Method: A Case Study in Korin Aquifer, Southeast Iran. *Journal of Spatial Hydrology* 3(1):1–5.

Lawes, J.B., J.C. Morton, J. Morton, J. Scott, and G. Thurber 1883. *The Soil of the Farm.* Orange Judd, Washington.

Lebbe, L., N. Van Meir, and P. Viaene. 2008. Potential implications of sea-level rise for Belgium. *J. Coastal Research* 24:358–366.

Lee, E., C. Seong, K. Hakkwan, S. Park, and M. Kang. 2010. Predicting the impacts of climate change on nonpoint source pollutant loads from agricultural small watershed using artificial neural network. *J Environ Sci* 22(6):840–845.

Lenderink, G., A. Buishand, and W. van Deursen. 2007. Estimates of future discharges of the river Rhine using two scenario methodologies: Direct versus delta approach. *Hydrol Earth Syst Sc* 11(3):1145–1159.

Leung, Y.K., K.H. Yeung, E.W.L. Ginn, and W.M. Leung. 2004. Global climate change: Cause of climate change by human activities. Hong Kong Observatory. No. 107.

Levavasseur, F., J.S. Bailly, P. Lagacherie, F. Colin, and M. Rabotin. 2012. Simulating the effects of spatial configurations of agricultural ditch drainage networks on surface runoff from agricultural catchments. *Hydrol Process* doi: 10.1002/hyp.8422.

Li, Y., B. Chen, Z. Wang, and S. Peng. 2011. Effects of temperature change on water discharge, and nutrient loading in the lower Pearl River basin based on SWAT modeling. *Hydrol Sci J* 56(1):68–83.

Limeira, J., L. Agullo, and M. Etxeberria. 2010. Dredged marine sand in concrete: An experimental section of a harbor pavement. *Constr Build Mater* 24(6):863–870.

Lindsay, W.L. 1979. *Chemical Equilibria in Soils.* John Wiley & Sons, New York, NY.

Lino, F.A.M., and K.A.R. Ismail. 2011. Energy and environmental potential of solid waste in Brazil. *Energ Policy* 39(6):3496–3502.

Liu, W., T. Wang, X. Gao, and Y. Su. 2004. Distribution and evolution of water chemical characteristics in Heihe River Basin. *J Desert Res* 24(6):755–762 ((in Chinese)).

Lloret, J., A. Marín, and L. Marín-Guirao. 2008. Is coastal lagoon eutrophication likely to be aggravated by global climate change? *Estuar Coast Shelf Sci* 78(2):403–412.

Loke, M.H., and R.D. Barker. 1996. Rapid least-squares inversion of apparent resistivity pseudosections by a quasi-Newton method. *Geophys Prospect* 44:131–152.

Lowrie, W. 2007. *Fundamentals of Geophysics.* Cambridge University Press. Cambridge, pp. 254–255.

Lufafa, A., M.M. Tenywa, M. Isabirye, M.J.G. Majaliwa, and P.L. Woomer. 2003. Prediction of soil erosion in alake victoria basin catchment using a GIS- based universal soil loss model. *Agricu Syst* 76:883–894.

Lukey, B.T., J. Sheffield, J.C. Bathurst, R.A. Hiley, and N. Mathys. 2000. Test of the SHETRAN technology for modelling the impact of reforestation on badlands runoff and sediment yield at Draix, France. *J Hydrol* 235:44–62.

Lyons, B.P., K.C. Pascoe, and R.B.I. McFadzen. 2002. Phototoxicity of pyrene and benzo[a] pyrene to embryo-larval stages of the pacific oyster crassostrea gigas. *Mar Environ Res* 54:627–631.

Ma, J.J., X.H. Sun, X.H. Guo, and Y.Y. Li. 2010b. Numerical simulation on soil water movement under water storage pits irrigation. *Transactions of the Chinese Society for Agricultural Machine* 41:46–51.

Macdonald, R.W., T. Harner, and J. Fyfe. 2005. Recent climate change in the Arctic and its impact on contaminant pathways and interpretation of temporal trend data. *Sci Total Environ* 342(1–3):5–86.

McKinney, D.C. 2015. Groundwater flow equations – Groundwater hydraulics presentation (available from this link: http://slideplayer.com/slide/2328551/).

McNamee, K., E. Wisheropp, C. Weinstein, A. Nugent, and L. Richmond. 2014. Scenario planning for building coastal resilience in the face of sea level rise: The case of Jacobs Avenue, Eureka, CA. *Humboldt J Soc Relat* 36:145–173.

Makwe, E., and C.D. Chup. 2013. Seasonal variation in physicochemical properties of groundwater around Karu abattoir. *EJESM* 6(5):489–497.

Mangor, K., N.K. Drønen, K.H. Kaergaard, and N.E. Kristensen. 2017. Shoreline management guidelines. DHI (www.dhigroup.com/marine-water/ebook-shoreline-man agement-guidelines).

Maroto-Valer, M.M., D.J. Fauth, M.E. Kuchta, Y. Zhang, and J.M. Andresen. 2005. Activation of magnesium rich minerals as carbonation feedstock materials for CO2 sequestration. *Fuel Process Technol* 86(14–15):1627–1645.

Marshall, E., and T. Randhir. 2008. Effect of climate change on watershed system: A regional analysis. *Clim Change* 89(3):263–280.

Martinez, M.L., A. Intralawan, G. Vazquez, O. Perez-Maqueo, P. Sutton, and R. Landgrave. 2007. The coasts of our world: Ecological, economic and social importance. *Ecol Econ* 63(2–3):254–272.

Martínková, M., C. Hesse, V. Krysanova, T. Vetter, and M. Hanel. 2011. Potential impact of climate change on nitrate load from the jizera catchment (Czech Republic). *Phys Chem Earth (A,B,C)* 36(13):673–683.

Matula, R.A. 1979. Electrical resistivity of copper, gold, palladium, and silver. *J Phys Chem Ref Data* 8(4):1147.

Meixner, T., A.H. Manning, D.A. Stonestrom et al. 2016. Implications of projected climate change for groundwater recharge in the western United States. *J Hydrol* 534:124–138.

Melillo, J.M., T.C. Richmond, and G.W. Yohe. 2014. Climate change impacts in the United States: The third national climate assessment. U.S. Global Change Research Program, pp 841. (NCA 3 report).

Ménard, C.B., J. Ikonen, K. Rautiainen, M. Aurela, A.N. Arslan, and J. Pulliainen. 2015. Effects of meteorological and ancillary data, temporal averaging, and evaluation methods on model performance and uncertainty in a land surface model. *J Hydrometeor* 16:2559–2576.

Mengel, D. 1995. *Roots, Growth and Nutrient Uptake*. Department of agronomy publication, Purdue University. West Lafayette, IN.

Merk, O., and T. Notteboom. 2013. The competitiveness of global port-cities: The case of Rotterdam, Amsterdam – The Netherlands. OECD Regional Development Working Papers, 2013/08, OECD Publishing.

Merrill, G.P. 1906. *A Treatise on Rocks, Rock-Weathering and Soils*. The MacMillan Company, New York, NY.

Mihaljević, M., and F. Stević. 2011. Cyanobacterial blooms in a temperate river-floodplain ecosystem: The importance of hydrological extremes. *Aquat Ecol* 45(3):335–349.

Mikutta, R., M. Kleber, M.S. Torn, and R. Jahn. 2006. Stabilization of soil organic matter: Association with minerals or chemical recalcitrance? *Biogeochemistry* 77:25–56.

Miller, W.D., and W.L. Harding. 2007. Climate forcing of the spring bloom in the Chesapeake Bay. *Mar Ecol Prog Ser* 331:11–22.

Millington, R.J., and J.M. Quirk. 1961. Permeability of porous solids. *Trans. Faraday Soc* 57:1200–1207.

Moosdorf, N., P. Renforth, and J. Hartmann. 2014. Carbon dioxide efficiency of terrestrial enhanced weathering. *Environ Sci Technol* 48:4809–4816.

Morris, D.A., and A.I. Johnson. 1967. Summary of hydrologic and physical properties of rock and soil materials as analyzed by the hydrologic laboratory of the U.S. Geological Survey 1948–1960. U.S. Geological Survey Water Supply Paper 1839-D. p. 42.

Mueller, B., and S.I. Seneviratne. 2012. Hot days induced by precipitation deficits at the global scale. *Proc. Natl. Acad. Sci. USA* 109:12 398–12 403.

Mukherjee, S., A. Mishra, and K.E. Trenberth. 2018. Climate change and drought: A perspective on drought indices. *Current Climate Change Reports* doi: org/10.1007/s40641-018-0098-x.

Munoz-Carpena, R., S. Shukla, and K. Shukla. 2004. Field devices for monitoring soil water content. Extension Bulletin, 343. Department of Agricultural and Biological Engineering, University of Florida, pp. 1–24.

Murray, R. 2008. The International Banking and Treating of Water in Aquifers. The lecture notes prepared by the department of water affairs and forestry, Pretoria, South Africa.

Nair, P.K.R., 2012. Climate change mitigation: A low-hanging fruit of agroforestry. In: *Agroforestry: The Future of Global Land Use*, Nair, P.K.R. and D. Garrity (Eds.). Springer, Dordrecht, pp. 31–67.
Nair, P.K.R., B.M. Kumar, and V.D. Nair. 2009. Agroforestry as a strategy for carbon sequestration. *J Plant Nutr Soil Sci* 172:10–23.
Neale, R.B., J. Richter, S. Park et al. 2013. The mean climate of the Community Atmosphere Model (CAM4) in forced SST and fully coupled experiments. *J Clim* 26:5150–5168.
Ng, G.H.C., D. McLaughlin, D. Entekhabi, and B.R. Scanlon. 2010. Probabilistic analysis of the effects of climate change on groundwater recharge. *Water Resour Res* 46:7502.
Niwas, S., and M. Celik. 2012. Equation estimation of porosity and hydraulic conductivity of Ruhrtal aquifer in Germany using near surface geophysics. *J Appl Geophys* 84:77–85.
Nõges, P., T. Nõges, M. Ghiani et al. 2011. Increased nutrient loading and rapid changes in phytoplankton expected with climate change in stratified South European lakes: Sensitivity of lakes with different trophic state and catchment properties. *Hydrobiologia* 66:255–270.
Nolan, P., J. O'Sullivan, and R. McGrath. 2017. Impacts of climate change on mid-twenty-first century rainfall in Ireland: A high-resolution regional climate model ensemble approach. *Int J Climatol* 37:4347–4363.
Nord, E.A., and J.P. Lynch. 2009. Plant phenology: A critical controller of soil resource acquisition. *J Exp Bot* 60:1927–1937.
O'Connor, W.K., D.C. Dahlin, D.N. Nilsen, R.P. Walters, and P.C. Turner. 2000. Carbon dioxide sequestration by direct mineral carbonation with carbonic acid. 25th International Technical Conference on Coal Utilization and Fuel Systems, Clearwater, FL, USA.
O'Connor, W.K., G.E. Rush, S.J. Gerdemann, L.R. Penner, and D.N. Nilsen. 2005. Aqueous mineral carbonation: Mineral availability, pretreatment, reaction parametrics and process studies, Albany Research Centre, USA DOE/ARC–TR-04-002.
O'Farrell, I., I. Izaguirre, G. Chaparro et al. 2011. Water level as the main driver of the alternation between a free-floating plant and a phytoplankton dominated state: A longterm study in a floodplain lake. *Aquat Sci Res Bound* 73(2):275–287.
Oates, K., and S.A. Barber. 1987. NUTRIENT UPTAKE: A microcomputer program to predict nutrient absorption from soil by roots. *Journal of Agronomic Education* 16(2):65–68.
Offodile, M.E. 1983. The occurrence and exploitation of groundwater in Nigeria basement complex. *J Mining Geol* 20:131–146.
Omosuyi, G.O., A. Adeyemo, and A.O. Adegoke. 2007. Investigation of groundwater prospect using electromagnetic and geoelectric sounding at afunbiowo, near Akure, Southwestern Nigeria. *Pacific J Sci Technol* 8:172–182.
Or, D., and M. Tuller. 2003. Hydraulic conductivity of unsaturated fractured porous media: Flow in a cross-section. *Adv Water Resour* 26:883–898.
Orwin, K.A., S.M. Buckland, D. Johnson et al. 2010. Linkages between plant traits and soil properties related to the functioning of temperate grassland. *J Ecol* 98:1074–1083.
Oseji, J.O., E.A. Atakpo, and E.C. Okolie. 2005. Geoelectric investigation of the aquifer characteristics and groundwater potential in Kwale, Delta State, Nigeria. *J Applied Sci Environ Mgt* 9:157–160.
Oulehle, F., and J. Hruska. 2009. Rising trends of dissolved organic matter in drinking-water reservoirs as a result of recovery from acidification in the Ore Mts., Czech Republic. *Environ Pollut* 157(12):3433–3439.

References

Özkan, K., E. Jeppesen, M. Søndergaard, T.L. Lauridsen, L. Liboriussen, and J. C. Svenning. 2012. Contrasting roles of water chemistry, lake morphology, land-use, climate and spatial processes in driving phytoplankton richness in the Danish landscape. *Hydrobiologia* 710(1):173–187.

Palmer, M.E., N.D. Yan, A.M. Paterson, and R.E. Girard. 2011. Water quality changes in south-central Ontario lakes and the role of local factors in regulating lake response to regional stressors. *Can J Fish Aquat Sci* 68:1038–1050.

Pandey, A., S. Himanshu, S. Mishra, and V. Singh. 2016. Physically based soil erosion and sediment yield models revisited. *Catena* 147:595–620.

Parker, B.R., D.R. Vinebrooke, and R.D. Schindler. 2008. Recent climate extremes alter alpine lake ecosystems. *Proc Natl Acad Sci* 105(35):12927–12931.

Parry, M.L. 2000. *Assessment of Potential Effects and Adaptations for Climate Change in Europe*. Jackson Environment Institute, University of East Anglia, Norwich.

Pawar, S.D., P. Murugavel, and D.M. Lal. 2009. Effect of relative humidity and sea level pressure on electrical conductivity of air over Indian Ocean. *J Geophys Res* 114 (D2):D02205.

Pervanchon, F., C. Bockstaller, and P. Girardin. 2002. Assessment of energy use in arable farming systems by means of an agro-ecological indicator: The energy indicator. *Agric Syst* 72:149–172.

Pierzynski, G.M., J.T. Sims, and G.F. Vance 2009. *Soils and Environmental Quality*. 3rd ed. CRC Press, Boca Raton, FL, US.

Pimentel, D., C. David, C.S. Rose et al. 2012. Annual vs. perennial grain production. *Agric Ecosyst Environ* 161:1–9.

Pitman, A.J. 2003. The evolution of, and revolution in, land surface schemes designed for climate models. *Int J Climatol* 23:479–510.

Pitz, C. 2016. Predicted impacts of climate change on groundwater resources of washington state environmental assessment program; Washington State Department of Ecology; Department of Ecology, State of Washington. Publication No. 16-03-006, 107p.

Pokrovsky, O.S., and J. Schott. 2000. Kinetics and mechanism of forsterite dissolution at 25 °C and pH from 1 to 12. *Geochim. Cosmochim. Acta* 64(2000):3313–3325.

Polyakov, V., and A. Fares. 2007. Evaluation of a non-point source pollution model, AnnAGNPS, in a tropical watershed. *Environ Model Software* 22(11):1617–1627.

Powlson, D.S., A. Bhogal, B.J. Chambers et al. 2012. The potential to increase soil carbon stocks through reduced tillage or organic additions: An England and Wales case study. *Agric Ecosyst Environ* 146:23–33.

Prathumratana, L., S. Sthiannopkao, and W.K. Kim. 2008. The relationship of climatic and hydrological parameters to surface water quality in the lower Mekong River. *Environ Int* 34(6):860–866.

Press, W., S. Flannery, S. Teukolsky, and W. Vetterling 1986. *Numerical Recipes: The Art of Scientific Computing*. Cambridge University Press. Cambridge.

Preston, B.L. 2004. Observed winter warming of the Chesapeake Bay estuary (1949–2002): Implications for ecosystem management. *Environ Manage* 34(1):125–139.

Ptashnyk, M. 2010. Derivation of a macroscopic model for nutrient uptake by hairy-roots, Nonlinear Analysis. *Real World Applications* 11:4586–4596.

Qian, B., E.G. Gregorich, S. Gameda, D.W. Hopkins, and X.L. Wang. 2011. Observed soil temperature trends associated with climate change in Canada. *J Geophys Res* 116:D2.

Quesney, A., S.L. Hégarat-Mascle, O. Taconet et al. 2000. Estimation of watershed soil moisture index from ERS/SAR data. *Remote Sens Environ* 72:290–303.

Raghunath, H.M. 2002. *Ground Water*. 2nd ed. New Age International Pvt. Ltd. New Delhi.

Rahman, S., and M.A. Rahman. 2015. Climate extremes and challenges to infrastructure development in coastal cities in Bangladesh. *Weather Climate Extremes* 7:84–95.

Rahmstorf, S. 2012. Modeling sea level rise. *Nature Education Knowledge* 3(10):4.

Rambal, S., J.M. Ourcival, R. Joffre et al. 2003. Drought controls over conductance and assimilation of a Mediterranean evergreen ecosystem: Scaling from leaf to canopy. *Global Change Biol* 9(12):1813–1824.

Rasmussen, T.C., K.G. Haborak, and M.H. Young. 2003. Estimating aquifer hydraulic properties using sinusoidal pumping at the Savannah River Site, South Carolina, USA. *Hydrogeology Journal* 11:466–482.

Rasse, D.P., C. Rumpel, and M.F. Dignac. 2005. Is soil carbon mostly root carbon? Mechanisms for a specific stabilization. *Plant Soil* 269:341–356.

Reich, P.B., S.E. Hobbie, T. Lee et al. 2006. Nitrogen limitation constrains sustainability of ecosystem response to CO2. *Nature* 440:922–925.

Ren, N., G. Yan, and J. Ma. 2006. The study on the influence of environmental factors of the submerged macrophytes in the East Lake. *Wuhan Univ. J Nat Sci Ed* 26(11):3594–3601.

Renard, K.G., G.R. Foster, G.A. Weesies, D.K. McCool, and D.C. Yoder 1997. *Predicting Soil Erosion by Water: A Guide to Conservation Planning with the Revised Universal Soil Loss Equation (RUSLE)*. USDA-ARS, Agricultural Handbook No. 703. Washington, DC.

Renforth, P. 2012. The potential of enhanced weathering in the UK. *Int J Greenhouse Gas Control* 10:229–243.

Reynolds, J.M. 1997. *An Introduction to Applied and Environmental Geophysics*. John Wiley & Sons, Chichester, UK.

Reynolds, S.G. 1970. The gravimetric method of soil moisture determination. *J Hydrol* 11(3):258–273.

Rice, T.J. 2003. Book review of: Soil Genesis and Classification, 5th ed. *Vadose Zone J* 2:767.

Rinck-Pfeiffer, S.M., S. Ragusa, P. Szajnbok, and T. Vandevelde. 2000. Interrelationships between biological, chemical and physical processes as an analog to clogging in aquifer storage and recovery (ASR) wells. *Water Res* 34:2110–2118.

Ritchey, E.L., J.M. McGrath, and D. Gehring 2015. *Determining Soil Texture by Feel*. Agriculture and Natural Resources Publications. Thailand, p. 139.

Ritzema, H.P. 1994. Subsurface flow to drains. In: *Drainage Principles and Applications, Publication 16*, H.P. Ritzema (Ed.), International Institute for Land Reclamation and Improvement (ILRI), Wageningen, The Netherlands, pp. 263–304. ISBN: 90 70753 3 39.

Rizzi, J., V. Gallina, S. Torresan, A. Critto, S. Gana, and A. Marceline. 2016. Regional risk assessment addressing the impacts of climate change in the coastal area of the Gulf of Gabes (Tunisia). *Sustain Sci* 11:455–476.

Roose, T., A.C. Fowler, and P.R. Darrah. 2001. A mathematical model of plant nutrient uptake. *J Math Biology* 42:347–360.

Roose, T., and G.J.D. Kirk. 2009. The solution of convection–Diffusion equations for solute transport to plant roots. *Plant Soil* 316:257–264.

Rosenzweig, C., G. Casassa, J.D. Karoly et al. 2007. Assessment of observed changes and responses in natural and managed systems. *Climate Change 2007: Impacts, Adaptation and Vulnerability. Contribution of Working Group II to the Fourth Assessment*

References

Report of the Intergovernmental Panel on Climate Change. Cambridge University Press, Cambridge, UK, 79–131.

Rounds, S.A., F.D. Wilde, and G.F. Ritz 2013. *Dissolved Oxygen (ver. 3.0): U.S. Geological Survey Techniques of Water-Resources Investigations.* book 9, chap. A6.2. U.S. Geological Survey. Reston, VA.

Rounsevell, M.D.A., and P.J. Loveland. 1992. An overview of hydrologically controlled soil responses to climate change in temperate regions. *SEESOIL* 8:69–78.

Roux, J.C. 1995. The evolution of groundwater quality in France: Perspectives for enduring use for human consumption. *Sci Total Environ* 171(1–3):3–16.

Ruffin, E. 1832. *An Essay on Calcareous Manures.* J.W. Campbell, Petersburg, VA.

Ruggiero, P. 1997. Wave runup on high energy dissipative beaches and the prediction of coastal erosion, Ph.D. thesis, Civil Engineering Department, Oregon State University, Corvallis, Oregon, p.14 5.

Salako Adebayo, O., and A. Adepelumi Abraham, 2018. Aquifer, classification and characterization. In: *Aquifers: Matrix and Fluids*, Javaid, M.S. and S.A. Khan (Eds.), IntechOpen. London, pp. 12–311, DOI: 10.5772/intechopen.72692.

Sallam, A., W.A. Jury, and J. Letey. 1984. Measurement of gas diffusion coefficient under relatively low air-filled porosity. *Soil Sci Soc Am J* 48:3–6.

Salvadori, G., and C. De Michele. 2004. Frequency analysis via copulas: Theoretical aspects and applications to hydrological events. *Water Resour Res* 40:W12511.

Samal, D., J.L. Kovar, B. Steingrobe, U.S. Sadana, P.S. Bhadoria, and N. Claassen. 2010. Potassium uptake efficiency and dynamics in the rhizosphere of maize (Zea mays L.), wheat (Triticum aestivum L.), and sugar beet (Beta vulgaris L.) evaluated with a mechanistic model. *Plant Soil* 332:105–121.

Sanchirico, J., and J. Wilen. 2005. Optimal spatial management of renewable resources: Matching policy scope to ecosystem scale. *J Environ Econ Manag* 50:23–46.

Sanderman, J., R. Farquharson, and J.A. Baldock. 2010. Soil carbon sequestration potential: A review for Australian agriculture. CSIRO sustainable agriculture flagship report, prepared for Department of Climate Change and Energy Efficiency. Available at: www.csiro.au/resources/Soil-Carbon-Sequestration-Potential-Report.html.

Santanello, J.A., C.D. Peters-Lidard, and S.V. Kumar. 2011. Diagnosing the sensitivity of local land–atmosphere coupling via the soil moisture-boundary layer interaction. *J Hydrometeorol* 12:766–786.

Saran, R.K., R. Kumar, and S. Yadav. 2017. Climate change: Mitigation strategy by various CO2 sequestration methods. *International Journal of Advanced Research in Science, Engineering* 6:299–308.

Sarkkola, S., H. Koivusalo, A. Laurén, P. Kortelainen, T. Mattsson, and M. Palviainen. 2009. Trends in hydrometeorological conditions and stream water organic carbon in boreal forested catchments. *Sci Total Environ* 408(1):92–101.

Schanz, T., W. Baille, and L. Ntuan. 2011. Effects of temperature on measurements of soil water content with time domain reflectometry. *Geotech Test J* 34(1):1–8.

Scharpenseel, H.W., A. Ayoub, and M. Schomaker 1990. *Soils on a Warmer Earth: Effects of Expected Climate Change on Soil Processes, with EMPhasis on the Tropics and Sub-tropics.* Elsevier, Amsterdam.

Schenk, H.J., and R.B. Jackson. 2002. Rooting depths, lateral root spreads and below-ground/above-ground allometries of plants in water-limited ecosystems. *J Ecology* 90:480–494.

Schenk, M.K., and S.A. Barber. 1980. Potassium and phosphorus uptake by corn genotypes grown in the field as influenced by root characteristics. *Plant Soil* 54:65–76.

Scheu, S., and J. Schauermann. 1994. Decomposition of roots and twigs: Effects of wood type (beech and ash), diameter, site of exposure and macrofauna exclusion. *Plant Soil* 241:155–176.

Schiedek, D., B. Sundelin, W.J. Readman, and W.R. Macdonald. 2007. Interactions between climate change and contaminants. *Mar Pollut Bull* 54(12):1845–1856.

Schnorbus, M., A. Werner, and K. Bennett. 2014. Impacts of climate change in three hydrologic regimes in British Columbia, Canada. *Hydrol Process* 28(3):1170–1189.

Schoeneberger, M., G. Bentrup, H. de Gooijer et al. 2012. Branching out: Agroforestry as a climate change mitigation and adaptation tool for agriculture. *J Soil Water Conserv* 67:128A–136A.

Schuur, E.A., J.G. Vogel, K.G. Crummer, H. Lee, J.O. Sickman, and T.E. Osterkamp. 2009. The effect of permafrost thaw on old carbon release and net carbon exchange from tundra. *Nature* 459(7246):556–559.

Seiller, G., and F. Anctil. 2014. Climate change impacts on the hydrologic regime of a Canadian river: Comparing uncertainties arising from climate natural variability and lumped hydrological model structures. *Hydrol Earth Syst Sc* 18:2033–2047.

Sellers, W.D. 1969. A global climatic model based on the energy balance of the earth atmosphere system. *J Appl Met* 8:392–400.

Senatore, A., G. Mendicino, G. Smiatek, and H. Kunstmann. 2011. Regional climate change projections and hydrological impact analysis for a Mediterranean basin in Southern Italy. *J Hydrol* 399(1-2):70–92.

Seneviratne, S.I., T. Corti, E.L. Davin et al. 2010. Investigating soil moisture–climate interactions in a changing climate: A review. *Earth Sci Rev* 99:125–161.

Seneviratne, S.I., D. Lüthi, M. Litschi, and C. Schär. 2006. Land–atmosphere coupling and climate change in Europe. *Nature* 443:205–209.

Shahid, S.A., M. Zaman, and L. Heng, 2018. Soil salinity: Historical perspectives and a world overview of the problem. In: *Guideline for Salinity Assessment, Mitigation and Adaptation Using Nuclear and Related Techniques*, Springer, Cham, pp. 43–53.

Shao, J., L. Li, Y. Cui, and Z. Zhang. 2013. Groundwater flow simulation and its application in groundwater resource evaluation in the North China plain, China. *Acta Geologica Sinica (English edition)* 87:243–253.

Sharad Jain, K., and V. Kumar. 2012. Trend analysis of rainfall and temperature data for India. *Curr Sci* 102(1):37–49.

Sharma, P.K., D. Kumar, H.S. Srivastava, and P. Patel. 2018. Assessment of different methods for soil moisture estimation: A review. *Journal of Remote Sensing and GIS* 9(1):57–73.

Shea, E.C. 2014. Adaptive management: The cornerstone of climate-smart agriculture. *J Soil and Water Conserv* 69(6):198A–199A.

Siham, K., B. Fabrice, A. Edine, and D. Patrick. 2008. Marine dredged sediments as new materials resource for road construction. *Waste Manage* 28(5):919–928.

Silliman, S.E., B.I. Borum, M. Boukari et al. 2010. Issues of sustainability of coastal groundwater resources: Benin, West Africa. *Sustainability* 2(8):2652–2675.

Singh, B.P., A.L. Cowie, and K.Y. Chan 2011. *Soil Health and Climate Change, Soil Biology*. Springer-Verlag, Berlin and Heidelberg, pp. 1–414.

Six, J., R.T. Conant, E.A. Paul, and K. Paustian. 2002. Stabilization mechanisms of soil organic matter: Implications for C-saturation of soils. *Plant Soil* 241:155–176.

Smith, G.D. 1978. *Numerical Solution of Partial Differential Equations: Finite Difference Methods*. Clarendon Press, Oxford.

References

Smith, P., D. Martino, Z. Cai et al. 2008. Greenhouse gas mitigation in agriculture. *Philosophical Transactions of the Royal Society B* 363:789–813.

Soil Survey Staff 2010. *Soil Taxonomy.* 11th ed. USDA National Resources Conservation Services, Washington DC.

Søndergaard, M., E.S. Larsen, B.T. Jørgensen, and E. Jeppesen. 2011. Using chlorophyll a and cyanobacteria in the ecological classification of lakes. *Ecol Indicators* 11(5):1403–1412.

Sonkamble, S., A. Sahya, N.C. Mondal, and P. Harikumar. 2012. Appraisal and evolution of hydrochemical processes from proximity basalt and granite areas of Deccan Volcanic Province (DVP) in India. *J Hydrol* 438-439:181–193.

Soomere, T., M. Viška, J. Lapinskis, and A. Räämet. 2011. Linking wave loads with the intensity of erosion along the coasts of Latvia. *Est J Eng* 17:359–374.

Steger-Hartmann, T., K. Kummerer, and J. Schecker. 1996. Trace analysis of the antineoplasics ifosfamide and cyclophosphamide in sewage water by two-step solid phase extraction and GS-MS. *J Chromat A* 726:179–184.

Steinbeiss, S., H. Beßler, C. Engels et al. 2008. Plant diversity positively affects short-term soil carbon storage in experimental grasslands. *Glob Change Biol* 14:2937–2949.

Sterrett, R.J. 2007. *Groundwater and Wells.* 3rd ed. Johnson Screens, New Brighton, MN.

Stockmann, U., B. Minasny, and A.B. McBratney. 2011. Quantifying processes of pedogenesis. *Adv Agron* 113(113):1–71.

Suddick, E.C., K.M. Scow, W.R. Horwath et al. 2010. The potential for California agricultural crop soils to reduce greenhouse gas emissions: A holistic evaluation. *Adv Agron* 107:123–162.

Sugita, F., and K. Nakane. 2007. Combined effects of rainfall pattern and porous media properties on nitrate leaching. *Vadose Zone J* 6:548–553.

Sun, X.H. 2002. Water storage pit irrigation and its functions of soil and water conservation. *J Soil and Water Conserv* 16:130–131.

Sun, Y., S. Solomon, A.G. Dai, and R.W. Portmann. 2007. How often will it rain?. *J Clim* 20:4801–4818.

Swanston Christopher, W., K. Janowiak Maria, A. Brandt Leslie et al. 2016. *Forest Adaptation Resources: Climate Change Tools and Approaches for Land Managers.* 2nd ed. Gen. Tech. Rep. NRS-GTR-87-2. U.S. Department of Agriculture, Forest Service, Northern Research Station, Newtown Square, PA. p. 161.

Swenson, J.J., C.E. Carter, J.C. Domec, and C.I. Delgado. 2011. Gold mining in the Peruvian amazon: Global prices, deforestation, and mercury imports. *PLoS One* 6:e18875.

Tadesse, N., K. Bheemalingeswara, and A.B.M. dulaziz. 2010. Hydrogeological investigation and groundwater potential assessment in haromaya watershed, eastern ethiopia. *Momona Ethiopian Journal of Science* 2:1.

Tate, E., J. Sutcliffe, D. Conway, and F. Farquharson. 2004. Water balance of Lake Victoria: Update to 2000 and climate change modelling to 2100/Bilan hydrologique du Lac Victoria: Mise à jour jusqu'en 2000 et modélisation des impacts du changement climatique jusqu'en 2100. *Hydrol Sci J* 49(4):563–574.

Taylor, C.J., and E.A. Greene, 2008. Hydrogeologic characterization and methods used in the investigation of karst hydrology. In: *Field Techniques for Estimating Water Fluxes between Surface Water and Ground Water*, Rosenberry, D.O. and J. W. LaBaugh (Eds.). U.S. Geological Survey, Reston, Virginia (EUA), pp. 71–114.

Tchounwou, P.B., C.G. Yedjou, A.K. Patlolla, and D.J. Sutton, 2012. Heavy metals toxicity and the environment. *EXS* 101:133–164.

Ternes, T.A., M. Stumpf, B. Scuppert et al. 1998. Simultaneous determination of antiseptics and acidic drugs in sewage and river water. *Wom Wasser* 90:295–309.

Thies, H., U. Nickus, V. Mair et al. 2007. Unexpected response of high alpine lake waters to climate warming. *Environ Sci Technol* 41(21):7424–7429.

Thodsen, H. 2007. The influence of climate change on stream flow in Danish rivers. *J Hydrol* 333:226–238.

Thomas, J. 1995. *Numerical Partial Differential Equations: Finite Difference Methods*. Springer-Verlag. New York.

Thomsen, D.C., T.F. Smith, and N. Keys. 2012. Adaptation or manipulation? Unpacking climate change response strategies. *Ecol Soc* 17(3):20.

To, J.P.C., J. Zhu, P.N. Benfey, and T. Elich. 2010. Optimizing root system architecture in biofuel crops for sustainable energy production and soil carbon sequestration. *F1000 Biol Rep* 2:65.

Toens, P., D. Visser, C. Van Der Westhuizen, W. Stander, and J. Rasmussen. 1998. Report on the overberg coastal groundwater resources, volume II. T and P Report, (980148).

Tonboe, R.T., S. Eastwood, T. Lavergne et al. 2016. The EUMETSAT sea ice concentration climate data record. *Cryosphere* 10:2275–2290.

Tong, S.T.Y., J.A. Liu, and A.J. Goodrich. 2007. Climate change impacts on nutrient and sediment loads in a Midewestern Agricultural Watershed. *J Environ Inf* 9(1):18–28.

Tõnisson, H., K. Orviku, J. Jaagus, Ü. Suursaar, A. Kont, and R. Rivis. 2008. Coastal damages on Saaremaa Island, Estonia, caused by the extreme storm and flooding on January 9, 2005. *J Coast Res* 24:602–614.

Tornevi, A., O. Bergstedt, and B. Forsberg. 2014. Precipitation effects on microbial pollution in a river: Lag structures and seasonal effect modification. *PLoS One* 9(5): e98546.

Toth, E., G. Gelybo, M. Dencso, I. Kasa, M. Birkas, and Á. Horel, 2018. Chapter 19 – Soil CO2 emissions in a long-term tillage treatment experiment A2 - Muñoz, María Ángeles. In: *Soil Management and Climate Change*, Zornoza, R. (Ed.). Academic Press. Cambridge, MA, pp. 293–307.

Trenberth, K.E. 1992. *Climate System Modeling*. Cambridge University Press. Cambridge, p. 787.

Trenberth, K.E., D.P. Jones, P. Ambenje et al. 2007. Observations: Surface and atmospheric climate change. In: Climate Change 2007: The Physical Science Basis. Contribution of Working Group I to the Fourth Assessment Report of the Intergovernmental Panel on Climate Change. Cambridge University Press, Cambridge, UK and New York, NY, pp. 235–336.

Trolle, D., D.P. Hamilton, C.A. Pilditch, I.C. Duggan, and E. Jeppesen. 2011. Predicting the effects of climate change on trophic status of three morphologically varying lakes: Implications for lake restoration and management. *Environ Model Softw* 26:354–370.

Tsang, Y.W., and C.F. Tsang. 1987. Channel model of flow through fractured media. *Water Resour Res* 23:467–479.

Tsutsumi, A., K. Jinno, and R. Berndtsson. 2004. Surface and subsurface water balance estimation by the groundwater recharge model and a 3-D two-phase flow model. *Hydrolog Sci J* 49(2):205–226.

Tu, J. 2009. Combined impact of climate and land use changes on streamflow and water quality in eastern Massachusetts, USA. *J Hydrol* 379(3–4):268–283.

Udagani, C. 2013. Gamma ray attenuation study with varying moisture content of clay bricks. *Int J Eng Sci Invent* 2(7):pp. 35–38.

References

USEPA. 2012. Guidelines for water reuse. EPA/600/R-12/618. National risk management research laboratory office of research and development, U.S. Agency for International Development, Cincinnati, OH, p. 643.

Van Biersel, T.P., D.A. Carlson, and L.R. Milner. 2007. Impact of hurricane storm surges on the groundwater resources. *Environ Geol* 53:813–826.

van Breemen, N., J. Mulder, and C.T. Driscoll. 1983. Acidification and alkalinization of soils. *Plant Soil* 75(3):283–308.

Van Dijk, J., M. Koenders, K. Rebel, M. Schaap, and M. Wassen. 2009. State of the art of the impact of climate change on environmental quality in the Netherlands. A framework for adaptation. Knowledge for climate. KfC report number KfC 006/09.

Van Loon, A.F., and H.A.J. Van Lanen. 2013. Making the distinction between water scarcity and drought using an observation-modeling framework. *Water Resour Res* 49:1483–1502.

van Vliet, M.T.H., W.H.P. Franssen, J.R. Yearsley et al. 2013. Global river discharge and water temperature under climate change. *Global Environ Chang* 23:450–464.

Varallyay, G. 2007. Potential impacts of climate change on agro-ecosystems. *Agric Conspectus Scient* 72:1–8.

Varallyay, G. 2010. The impact of climate change on soils and on their water management. *Agronomy Research* 8(Special Issue II):385–396.

Verhaar, M.P., M.P. Biron, I.R. Ferguson, and B.T. Hoey. 2011. Implications of climate change in the twenty-first century for simulated magnitude and frequency of bed-material transport in tributaries of the Saint-Lawrence River. *Hydrol. Process.* 25:1558–1573.

Veríssimo, H., M. Lane, J. Patrício, S. Gamito, and C.J. Marques. 2013. Trends in water quality and subtidal benthic communities in a temperate estuary: Is the response to restoration efforts hidden by climate variability and the estuarine quality paradox?. *Ecol Indicators* 24:56–67.

Vermeer, M., and S. Rahmstorf. 2009. Global sea level linked to global temperature. *Proc Natl Acad Sci USA* 106:21527–21532.

Vestergaard, P. 1991. Morphology and vegetation of a dune system in SE Denmark in relation to climate change and sea-level rise. *Landsc Ecol* 6:77–87.

Visbeck, M., U. Kronfeld-Goharani, B. Neumann et al. 2014. Securing blue wealth: The need for a special sustainable development goal for the ocean and coasts. *Mar Policy* 48:184–191.

Viviroli, D., and R. Weingartner. 2004. The hydrological signifance of mountains: From regional to global scale. *Hydrology and Earth System Sciences* 8(6):1016–1029.

Waibel, M.S., M.W. Gannett, H. Chang, and C.L. Hulbe. 2013. Spatial variability of the response to climate change in regional groundwater systems – Examples from simulations in the Deschutes Basin, Oregon. *J Hydrol* 486:187–201.

Wallace, J., L. Stewart, A. Hawdon, R. Keen, F. Karim, and J. Kemei. 2009. Flood water quality and marine sediment and nutrient loads from the Tully and Murray catchment in north Queensland. *Austrilia. Mar Freshw Res* 60(11):1123–1131.

Walthall, C., J. Hatfield, P. Backlund et al. 2012. *Climate Change and Agriculture in the United States: Effects and Adaptation.* U.S. Department of Agriculture, Washington, DC. p. 186.

Wan, Y., E. Lin, W. Xiong, Y. Li, and L. Guo. 2011. Modeling the impact of climate change on soil organic carbon stock in upland soils in the 21st century in China. *Agric Ecosyst Environ* 141:23–31.

Wang, S., X. Jin, Q. Bu, L. Jiao, and F. Wu. 2008. Effects of dissolved oxygen supply level on phosphorus release from lake sediments. *Coll. Surf. A: Physicochemical Eng Aspects* 316(1):245–252.

Wang, X., and M.M. Maroto-Valer. 2011. Dissolution of serpentine using recyclable ammonium salts for CO2 mineral carbonation. *Fuel* 90(3):1229–1237.

Wang, X., M. Yang, X. Liang et al. 2014. The dramatic climate warming in the Qaidam Basin, northeastern Tibetan Plateau, during 1961–2010. *IJCli* 34(5):1524–1537.

Washington, W.M., and C.L. Parkinson 2005. *An Introduction to Three-dimensional Climate Modeling*. 2nd ed. University Science Books. Sausalito, CA, p. 353.

Watts, C.W., W.R. Whalley, P.C. Brookes, B.J. Devonshire, and A.P. Whitmore. 2005. Biological and physical processes that mediate micro-aggregation of clays. *Soil Sci* 170:573–583.

Weigel, A.P., M.A. Liniger, C. Appenzeller, and R. Knutti. 2010. Risks of model weighting in multi-model climate projections. *J Clim* 23(15):4175–4191.

Wernberg, T., D.A. Smale, F. Tuya et al. 2013. An extreme climatic event alters marine ecosystem structure in a global biodiversity hotspot. *Nat Clim Change* 3(1):78–82.

White, R.E. 1979. *Introduction to the Principles and Practice of Soil Science*. Blackwell, Oxford.

Whitehead, P., A. Wade, and D. Butterfield. 2009a. Potential impacts of climate change on water quality in six UK rivers. *Hydrol Res* 40(2-3):113–122.

Widory, D., W. Kloppmann, L. Chery, J. Bonnin, H. Rochdi, and J.L. Guinamant. 2004. Nitrate in groundwater: An isotopic multi-tracer approach. *J Contam Hydrol* 72 (1-4):165–188.

Wiedner, C., J. Rücker, R. Brüggemann, and B. Nixdorf. 2007. Climate change affects timing and size of populations of an invasive cyanobacterium in temperate regions. *Oecologia* 152(3):473–484.

Wilby, R., P. Whitehead, J.A. Wade, D. Butterfield, R.J. Davis, and G. Watts. 2006. Integrated modelling of climate change impacts on water resources and quality in a lowland catchment: River Kennet, UK. *J Hydrol* 330(1-2):204–220.

Wild, A. 1993. *Soils and the Environment*. Cambridge University Press, Cambridge, UK.

Wilhelm, S., and R. Adrian. 2008. Impact of summer warming on the thermal characteristics of a polymictic lake and consequences for oxygen. *Nutrients and Phytoplankton. Freshw Biol* 53(2):226–237.

Wilks, D.S. 1992. Adapting stochastic weather generation algorithms for climate change studies. *Clim Change* 22(1):67–84.

Wilson, C.O., and Q. Weng. 2011. Simulating the impacts of future land use and climate changes on surface water quality in the Des Plaines River watershed, Chicago Metropolitan Statistical Area, Illinois. *Sci Total Environ* 409(20):4387–4405.

Winter, T.C., J.W. Harvey, O.L. Franke, and W.M. Alley 1998. *Ground Water and Surface Water; a Single Resource*. USGS Circular 1139. U.S. Geological Survey, Denver, Colorado.

Winterdahl, M., H. Laudon, S.W. Lyon, C. Pers, and K. Bishop. 2016. Sensitivity of stream dissolved organic carbon to temperature and discharge: Implications of future climates. *J Geophys Res Biogeosci* 121:126–144.

Wintgens, T., R. Hochstrat, C. Kazner, P. Jeffrey, B. Jefferson, and T. Melin, 2012. Managed aquifer recharge as a component of sustainable water strategies-a brief guidance for EU policies. In: *Water Reclamation Technologies for Safe Managed Aquifer Recharge*, Kazner, C., T. Wintgens and P. Dillon (Eds.). IWA Publishing, London, pp. 411–429.

References

Wischmeier, W.H., and D.D. Smith 1965. *Predicting Rainfall-erosion Losses from Cropland East of the Rocky Mountains: Guide for Selection of Practices for Soil and Water Conservation.* US Department of Agriculture (USDA). Washington, DC, No. 282.

Wolfe, D., J. Beem-Miller, L. Chambliss, A. Chatrchyan, and H. Menninger, 2011. Farming success in an uncertain climate. In: *Adaptation Resources for Agriculture: Responding to Climate Variability and Change in the Midwest and Northeast*, Extension, C.C. (Ed.). Cornell University, Ithaca, NY, p. 4.

Worrall, F., A. Armstrong, and J. Holden. 2007. Short-term impact of peat drain-blocking on watercolour, dissolved organic carbon concentration, and water table depth. *J Hydrol* 337(3-4):315–325.

Wu, Q., and X.H. Xia. 2014. Trends of water quantity and water quality of the Yellow River from 1956 to 2009: Implication for the effect of climate change. *Environmental System Research* 3(1):1–6.

WWAP. 2009. The United Nations World Water Development Report 3, water in a changing world, world water assessment programme. Paris, UNESCO Publishing, p. 349.

Xia, X.H., Z.F. Yang, and X.Q. Zhang. 2009. Effect of suspended sediment concentration on nitrification in river water: Importance of suspended sediment–water interface. *Environ Sci Technol* 43:3681–3687.

Xia, X.H., Z.H. Yang, G.H. Huang, X.Q. Zhang, H. Yu, and X. Rong. 2004. Nitrification in natural waters with high suspended solid content—A study for the Yellow River. *Chemosphere* 57:1017–1029.

Xia, X.H., H. Yu, Z.F. Yang, and G.H. Huang. 2006. Biodegradation of polycyclic aromatic hydrocarbons in the natural waters of the Yellow River: Effects of high sediment content on biodegradetion. *Chemosphere* 65:457–466.

Xu, Y., Q. Cai, L. Ye, and M. Shao. 2010. Asynchrony of spring phytoplankton response to temperature driver within a spatial heterogeneity bay of Three-Gorges Reservoir, China. *Limnol Ecol Manage Inland Waters* 41:174–180.

Ylhaisi, J.S., H. Tietavainen, P. Peltonen-Sainio et al. 2010. Growing season precipitation in Finland under recent and projected climate. *Nat Hazards Earth Syst Sci* 10(7):1563–1574.

Young, A. 1997. *Agroforestry for Soil Management.* C.A.B. International and ICRAF, Wallingford, UK.

Young, F.J. 1988. *Soil Survey of Territory of Guam.* Soil Conservation Service, U.S. Department of Agriculture, Washington DC. p. 166.

Younis, J., S. Anquetin, and J. Thielen. 2008. The benefit of high-resolution operational weather forecasts for flash-flood warning. *Hydrol Earth Syst Sci* 5:345–377.

Yu, J.T., E.J. Bouwer, and M. Coelhan. 2006. Occurrence and biodegradability studies of selected pharmaceuticals and personal care products in sewage effluent. *Agric Water Manag* 86:72–80.

Yuan, Z.Y., and L.H. Li. 2007. Soil water status influences plant nitrogen use: A case study. *Plant Soil* 301:303–313.

Yunev, O.A., J. Carstensen, S. Moncheva, A. Khauliulin, G. Ertebjerg, and S. Nixon. 2007. Nutrient and phytoplankton trends on the western Black Sea shelf in response to cultural eutrophication and climate changes. *Estuar Coast Shelf Sci* 74(1-2):63–76.

Zaehle, S., P. Friedlingstein, and A.D. Friend. 2010. Terrestrial nitrogen feedbacks may accelerate future climate change. *Geophys Res Lett* 37:L01401.

Zevenhoven, R., and J. Kohlmann. 2002. CO2 sequestration by magnesium silicate mineral carbonation in Finland, recovery, recycling and re-integration, Geneva, Switzerland.

Zevenhoven, R., and S. Teir. 2004. Long-term storage of CO2 as magnesium carbonate in Finland, 3rd Annual Conference on Carbon Capture and Sequestration, Alexandria, VA, USA.

Zhang, J., W.C. Wang, and L. Wu. 2009. Land-atmosphere coupling and diurnal temperature range over the contiguous United States. *Geophys Res Lett* 36:L06706.

Zhang, J.L., T.J. Flowers, and S.M. Wang. 2010. Mechanisms of sodium uptake by roots of higher plants. *Plant Soil* 326:45–60.

Zhang, K., B. Douglas, and S. Leatherman. 2000. Do storms cause long-term erosion along the U.S. east barrier coast?. *J Geol* 110(4):493–502.

Zhang, L., W. Lu, Y. An, D. Li, and L. Gong. 2012. Response of non-point source pollutant loads to climate change in the Shitoukoumen reservoir catchment. *Environ Monit Assess* 184:581–594.

Zhang, Y., W. Chen, S.L. Smith, D.W. Riseborough, and J. Cihlar. 2005. Soil temperature in Canada during the twentieth century: Complex responses to atmospheric climate change. *J Geophys Res* 110:D03112.

Zhou, J., J.Y. Lin, S.H. Cui, Q.Y. Qiu, and Q.J. Zhao. 2013. Exploring the relationship between urban transportation energy consumption and transition of settlement morphology: A case study on Xiamen Island, China. *Habitat Int* 37:70–79.

Zhu, Y.M., L.J. Zhang, and Y. Wu. 1997. Back analysis of coefficient tensor of permeability for fractured rock mass. *Chinese Journal of Rock Mechanics and Engineering* 16(1):461–470.

Zodhy, A.A., G.P. Eaton, and D.R. Mayhey, 1974. Applications of surface geophysics to groundwater investigations. In: *Techniques of Water Resources Investigations of the U.S. Geological Survey*, A.A.R. Zohdy, G.P. Eaton, and D.R. Mabey (Eds.). Book 2, Chapter D1, p. 116.

Zwiener, C., S. Seeger, T. Glauner et al. 2002. Metabolites from the biodegradation of pharmaceutical residues of ibuprofen in biofilm reactor and batch experiment. *Anal Bioanal Chem* 372:569–575.

Index

C

climate modeling, 83
    atmospheric general circulation model, 84, 86
    land model, 83, 87
    ocean model, 83, 86, 87
    sea ice model, 83, 89
coastal aquifers, 98, 100, 110, 111, 112, 113
    aquifer properties, 61, 104
    aquifer recharge, 110, 111, 112, 134, 138, 139, 140, 142, 143, 144, 145, 146, 147
    confined aquifers, 98, 99, 104, 141, 142
    saltwater intrusion, 59, 66, 100, 110, 111, 112, 138
    unconfined aquifers, 98, 99, 104
coastal erosion, 115, 116, 117, 118, 119, 120, 121, 122, 123, 147
    erosion modeling, 120
    human-induced factors, 115, 119, 122, 123
    natural factors, 115, 123

G

global climate change, 13, 35, 37, 39, 41
    droughts, 6, 35, 42, 53, 59, 97, 125, 126, 127, 151
    floods, 35, 41, 42, 97, 110, 125, 126, 127, 139, 151
    increasing air temperature, 35
    ocean acidification, 35, 40, 41, 76, 151
    rainfall patterns change, 38, 50
    sea level rise, 35, 39, 40, 59, 65, 66, 97, 100, 110, 111, 112, 115, 118, 122, 133, 147, 151

M

management strategies, 42, 43, 133, 134, 135, 137, 139, 141, 143, 145, 147, 149, 150

S

soil carbon sequestration, 19, 20, 67, 81
soil erosion, 9, 49, 51, 52, 59, 78, 79, 116, 120, 121, 126, 133, 134, 135, 147, 148, 151
soil processes, 20, 23, 25, 26, 27, 29, 30, 31, 33, 48

W

water resources, 16, 39, 42, 43, 55, 56, 57, 59, 61, 63, 65, 66, 97, 98, 99, 101, 103, 105, 107, 109, 110, 111, 113, 125, 126, 127, 129, 131, 133, 137, 138, 139, 150, 151

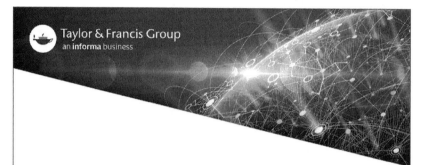